编委会

主 编
俞汉青

编 委
（以姓氏拼音排序）

陈洁洁	刘 畅	刘武军
刘贤伟	卢 姝	吕振婷
裴丹妮	盛国平	孙 敏
汪雯岚	王楚亚	王龙飞
王维康	王允坤	徐 娟
俞汉青	虞盛松	院士杰
翟林峰	张爱勇	张 锋

"十四五"国家重点出版物出版规划重大工程

污染控制理论与
应用前沿丛书

环境功能二氧化钛纳米材料 降解污染物及其催化机理研究

TiO₂-based Nanomaterials with Environmental Function
for Improving Pollutants Degradation
and Exploring Its Catalytic Mechanism

王维康　著
龙珺璐

中国科学技术大学出版社

内容简介

二氧化钛（TiO_2）基半导体纳米材料作为一种高效、稳定且环境友好的光催化剂，在环境领域有着广泛应用。然而其可见光利用率低、电子空穴复合严重和催化机理不清晰等问题，严重制约了其在水污染治理领域的实际应用。本书针对上述三个问题，从 TiO_2 基材料晶体结构、能带结构入手，通过高活性晶面合成、元素掺杂、晶格缺陷引入、相结/异质结构筑等调控手段，使改性后的 TiO_2 基纳米材料的光催化性能显著提升，实现了典型难降解有机污染物（如双酚 A、阿特拉津、罗丹明 B 等）的高效降解。以此为基础，利用荧光寿命成像系统探索 TiO_2 基光催化剂在单分子单颗粒水平上的微观催化机理，为发挥其在污染物降解中的作用提供科学依据，也为其在水污染控制领域的实际应用提供理论指导。

图书在版编目(CIP)数据

环境功能二氧化钛纳米材料降解污染物及其催化机理研究/王维康，龙琭璐著. —合肥：中国科学技术大学出版社，2022.3
（污染控制理论与应用前沿丛书/俞汉青主编）
国家出版基金项目
"十四五"国家重点出版物出版规划重大工程
ISBN 978-7-312-05392-4

Ⅰ.环… Ⅱ.①王… ②龙… Ⅲ.二氧化钛—纳米材料—应用—污染物—光降解—研究 Ⅳ.X5

中国版本图书馆 CIP 数据核字(2022)第 031458 号

环境功能二氧化钛纳米材料降解污染物及其催化机理研究

HUANJING GONGNENG ERYANGHUATAI NAMI CAILIAO JIANGJIE WURANWU JI QI CUIHUA JILI YANJIU

出版	中国科学技术大学出版社
	安徽省合肥市金寨路96号,230026
	http://www.press.ustc.edu.cn
	https://zgkxjsdxcbs.tmall.com
印刷	安徽联众印刷有限公司
发行	中国科学技术大学出版社
开本	787 mm×1092 mm 1/16
印张	14.5
字数	276 千
版次	2022 年 3 月第 1 版
印次	2022 年 3 月第 1 次印刷
定价	90.00 元

总　序

建设生态文明是关系人民福祉、关乎民族未来的长远大计,在党的十八大以来被提升到突出的战略地位。2017 年 10 月,党的十九大报告明确提出"污染防治"是生态文明建设的重要战略部署,是我国决胜全面建成小康社会的三大攻坚战之一。2018 年,国务院政府工作报告进一步强调要打好"污染防治攻坚战",确保生态环境质量总体改善。这都显示出党和国家推动我国生态环境保护水平同全面建成小康社会目标相适应的决心。

当前,我国环境污染状况有所缓解,但总体形势仍然严峻,已严重制约了我国经济社会的持续健康发展。发展以资源回收利用为导向的污染控制新理论与新技术,是进一步推动污染物高效、低成本、稳定去除的发展方向,已成为国家重大战略需求和国际重要学术前沿。

为了配合国家对生态文明建设、"污染防治攻坚战"的一系列重大布局,抢占污染控制领域国际学术前沿制高点,加快传播与普及生态环境污染控制的前沿科学研究成果,促进相关领域人才培养,推动科技进步及成果转化,我们组织一批来自多个"双一流"大学、活跃在我国环境科学与工程前沿领域、有影响力的科学家共同撰写"污染控制理论与应用前沿丛书"。

本丛书是作者团队承担的国家重大重点科研项目(国家重大科技专项、国家 863 计划、国家自然科学基金)和获得的重大科技成果奖励(2014 年国家自然科学奖二等奖、2020 年国家科学技术进步奖二等奖)的系统总结,是作者团队攻读博士学位期间取得的重要的前沿学术成果(全国百篇优秀博士论文、中科院优秀博士论文等)的系统凝练,是一套系统反映污染控制基础科学理论与前沿高新技术研究成果的系列图书。本丛书围绕我国环境领域的污染物生化控制、转化机理、无害化处置、资源回收利用等亟须解决的一些重大科学问题与技术问题,将物理学、化学、生物学、材料学等学科的最新理论成果以及前沿高新技术应用到污染控制过程中,总结了我国目前在污染控制领域(特别是废水和固废领域)的重要研究进展,探索、建立并发展了常温空气阴极燃料电池、纳米材料、新兴生物电化学系统、新型膜生物反应器、水体污染物的化学及生物转化,以及固体废弃物污染控制与清洁转化等方面的前沿理论与技术,形成了具有广阔应用前景的新理论和新方法,为污染控制与治理提供了理论基础和科学依据。

"污染控制理论与应用前沿丛书"是服务国家重大战略需求、推动生态文明建设、打赢"污染防治攻坚战"的一套丛书。其出版将有利于促进最前沿的科研成果得到及时的传播和应用,有利于促进污染治理人才和高水平创新团队的培养,有利于推动我国环境污染控制和治理相关领域的发展和国际竞争力的提升;同时为环境污染控制与治理实践提供新思路、新技术、新材料,也可以为政府环境决策、强化环境管理、履行国际环境公约等提供科学依据和技术支撑,在保障生态环境安全、实施生态文明建设、打赢"污染防治攻坚战"中起到不可替代的作用。

编委会

2021 年 10 月

前　言

光催化是一种能直接利用自然界中的太阳光实现污染物催化净化的高级氧化技术。光催化剂的性能是决定该技术的主控因素。因此，如何设计、合成、优化高性能光催化剂，并探究其微观催化机制，是该领域的重要研究问题。二氧化钛（TiO_2）作为一种高效、稳定且环境友好的宽带隙光催化剂，仍存在暴露活性晶面低、无可见光吸收和量子效率低等问题。本书针对上述问题，以典型的光催化剂 TiO_2 为对象，开展设计、制备与优化、反应过程机理解析和污染物净化应用效果评价等研究工作，力图通过简单温和的合成方法制备高效的 TiO_2 催化剂，建立实现宽带隙催化剂良好可见光响应和强化光催化性能的有效手段，发展提高窄带隙光催化电子空穴分离效率的新策略，进而提高光催化剂的适用范围和催化效率，为其实际应用提供理论指导和技术支撑。

针对特定反应活性晶面暴露低的问题，以铂（Pt）负载高暴露{001}氧化晶面的锐钛矿 TiO_2 为研究对象，通过理论计算，发现 Pt 负载高暴露{001}氧化晶面 TiO_2 能够提高光催化降解硝基苯的速率。在理论分析指导上，采用简捷、绿色的合成路线制备了{001}氧化晶面高暴露且均匀负载 Pt 纳米颗粒的十面体 TiO_2 催化剂，用于有机污染物——硝基苯的光催化降解。实验表明，与将 Pt 纳米颗粒负载在{101}氧化晶面的 TiO_2 相比，均匀负载 Pt 纳米颗粒的高暴露{001}氧化晶面的 TiO_2，可以显著提升硝基苯的光降解效率和光转化效率。考虑到相应的{101}作为还原活性晶面，可能针对还原反应具有更好的催化效果，通过一步水热反应合成了非化学计量的锐钛矿 TiO_2 微球，其具有高暴露的{101}还原晶面。实验结果证明，该材料表现出很高的光催化还原活性，与常规的锐钛矿相比，其氢气产生速率明显提高。锐钛矿 TiO_2 的{101}还原晶面能够传递更多的光生电子，从而促进光催化产氢；通过飞秒时间分辨瞬态吸收光谱分析，发现自掺杂的 Ti^{3+} 可以作为活性位点捕获电子，同转移到{101}还原晶面的电子共同强化其光催化还原作用。为进一步提高暴露活性晶面，开展了介晶结构和高能晶面暴露 TiO_2 微米棒的研究工作。利用阳极氧化和煅烧结合的方法在碳片表面制备了新颖的六角形 TiO_2 微米棒介晶结构。此微米棒是以一种中间体支架为支撑，通过向中间体空间内填充 TiO_2 单晶小颗粒定向自组装形成的。TiO_2 的晶面暴露情况、结晶性以及结构特点均对 TiO_2 的电子空穴分离效率和光催化性能有显著影响。

进一步为了使高暴露活性晶面的 TiO_2 纳米材料具有可见光活性，利用掺杂新方法来改性并提升 TiO_2 单晶可见光光催化活性。以废弃的电解液为原料，以 6 种具有代表性 3 类异质元素（金属元素、非金属元素和稀土金属元素）为掺杂对象，发展了一种绿色的各类异质元素均可以实现掺杂的通用方法，获得了结晶性好、形态规则、暴露{001}晶面晶格缺陷少且异质元素掺杂的 TiO_2 单晶。结果表明，此方法能够将异质元素掺杂进入 TiO_2 单晶中，而不明显破坏 TiO_2 单晶的形貌；在异质元素掺杂之后，TiO_2 单晶的电化学和光化学性能都有了明显提高。考虑到金红石 TiO_2 相相比于锐钛矿 TiO_2 相具有更好的光的吸收，同时两相的组合也有利于光生电子空穴的有效分离；由此，以金红石/锐钛矿两相的相结 TiO_2 催化剂作为研究对象，利用一步煅烧法，调控原料中 Ti^{3+} 与 HCl 的比例，在低于相变温度下制得金红石/锐钛矿比例可控的相结 TiO_2 催化剂，并应用于有机污染物的光催化降解和氢气的产生。结果显示，比例优化的金红石/锐钛矿相结实现了光生载流子的有效分离和转移，形成了有效的活性催化界面，从而促进光催化过程中活性氧化物种的生成，实现了环境有机污染物的高效降解。因此，我们考虑通过结合相结与掺杂的双重调控，实现更高的 TiO_2 光催化性能。在前述相结的基础上，加入硼酸作为前驱体，在相同的煅烧方法下，合成了一种硼掺杂调控金红石/锐钛矿比例的相结 TiO_2，并用于光催化降解阿特拉津。该材料在金红石/锐钛矿相结的基础上，掺杂的硼原子则作为电子的捕获位点，能够更有效地介导电子的传递；所制备的硼掺杂的相结 TiO_2 表现出更高的光催化降解阿特拉津的能力，其降解速率远远高于未掺杂催化剂的；并通过对光催化活性物种以及反应中间产物的分析，阐述了反应过程的机理。

为有效提升 TiO_2 基纳米光催化剂的可见光吸收及促进电子空穴的有效分离，可以通过与窄带隙光催化剂构筑高活性的异质结，实现光催化有机物污染物的高效降解。过渡金属硫化物作为典型窄带隙光催化剂，以 CdS 为例，我们提出了直接利用 ZnO 中低还原性的导带电子作为固体牺牲试剂，有效地从 CdS 中捕获低氧化能的价带空穴，实现高效催化降解难以用生化方法去除的有机污染物。结果证明，制备的 $ZnO/CdS/TiO_2$ 复合材料在太阳光照射下，无需任何额外的牺牲试剂，对目标污染物 RhB 和阿特拉津的降解以及纺织废水的处理，比典型的 CdS/TiO_2 类似物具有更高的光催化活性。此外，在模拟太阳光照射下，ZnO 的负载，使 $ZnO/CdS/TiO_2$ 复合材料在水溶液光催化反应中的光腐蚀大大降低，表观抗光腐蚀能力明显提高。这种新型的 $ZnO/CdS/TiO_2$ 异质结的构筑，不仅提高了催化活性，还解决了 CdS 的稳定性问题，为容易光腐蚀催化材料的设计与应用提供了一种新的解决思路，并增加了其实际应用的可能性。考虑到硫化物对水体本身存在二次污染的影响，希望通过利用稳定的非金属半导体作为复合材料用于提高 TiO_2 的可见光光催化活性。因此，我们成功地构建了 Ti—N 和 C—O 双键桥连的二维（2D）TiO_2-g-C_3N_4，并负载在碳纸上构筑成电极，实现了高效的光电化学（PEC）污染物降解。密度泛函理论（DFT）计算表明，生成的 2D TiO_2-g-C_3N_4 的界面异质结形成有效的内电场，并且通过 Ti—N 和 C—O 双键实现快速的电荷分离和转移。此外，根据电化学阻抗谱，与单一的 TiO_2 或 g-C_3N_4 相比，界面异质结的 2D TiO_2-g-C_3N_4 复合材料明显降低了其内阻，更有利于电催化。结果表明，2D TiO_2-g-C_3N_4 碳纸电极展现了更高的 PEC 活性，实现了对典型难降解污染物双酚 A（BPA）的去除。同时，光电催化系统中最大的 BPA 降解率远高于光催化和电催化系统的总和，实现了"1+1>2"的效果。

为了更为直观地探究 TiO_2 的催化活性位点与催化机理

分析，我们首先利用共聚焦显微镜与时间分辨仪进行单颗粒 TiO$_2$ 的时间空间分辨荧光寿命成像研究，发现在 {101} 和 {001} 晶面之间处荧光强度最高，且电子寿命最长。然后，选取中性红和刃天青作为氧化还原染料，利用荧光显微镜研究单颗粒 TiO$_2$ 上的光催化染料分子荧光变化，实现了单颗粒 TiO$_2$ 空间分辨的荧光测定，并发现单颗粒 TiO$_2$ 的氧化还原反应位点都位于 {101} 和 {001} 晶面之间，与单颗粒荧光寿命成像结果一致。该结果不仅提供了 TiO$_2$ 光催化反应微观机理研究的原位观测技术，而且可以拓展应用于非均相 B 催化系统和生物系统中氧化还原反应的实时观测，为环境催化领域的机理研究提供了一种新的技术手段。

作 者

2021 年 10 月

目 录

总序 —— i

前言 —— iii

第 1 章
绪论 —— 001

1.1 光催化的定义及二氧化钛材料的研究进展
　　—— 003

1.2 二氧化钛在环境催化领域中的应用 —— 007

1.3 二氧化钛的晶体类型及催化机制 —— 012

1.4 二氧化钛在催化过程中的原位/微观研究
　　—— 016

第 2 章
铂负载高暴露{001}晶面的锐钛矿二氧化钛合成及其光催化性能 —— 035

2.1 高暴露{001}晶面的锐钛矿二氧化钛研究进展
　　—— 037

2.2 铂负载高暴露{001}晶面的锐钛矿二氧化钛的合成与表征 —— 039

2.3 铂负载高暴露{001}晶面的锐钛矿二氧化钛的光催化硝基苯机制分析 —— 050

第 3 章
{001}晶面刻蚀蓝色锐钛矿二氧化钛微米球的合成及其光催化性能 —— 059

3.1 高暴露{101}晶面的含氧空位锐钛矿二氧化钛研究进展 —— 061

3.2 晶面刻蚀蓝色锐钛矿二氧化钛的合成与表征
　　—— 062

3.3 晶面刻蚀蓝色锐钛矿二氧化钛的光催化产氢性能
　　—— 070

3.4 晶面刻蚀蓝色锐钛矿二氧化钛的光催化机制分析 —— 072

第 4 章
{001}二氧化钛介晶结构自生长过程及其光催化性能的研究 —— 079

4.1 {001}二氧化钛介晶结构自生长研究进展 —— 081

4.2 {001}二氧化钛介晶的合成和表征 —— 082

4.3 {001}二氧化钛介晶的形成分析 —— 084

4.4 {001}二氧化钛介晶的物相分析 —— 086

4.5 {001}二氧化钛介晶的光催化性能研究 —— 088

第 5 章
二氧化钛纳米单晶通用元素掺杂改性方法及光催化性能的研究 —— 095

5.1 二氧化钛纳米单晶掺杂改性研究进展 —— 097

5.2 掺杂二氧化钛单晶的合成与表征 —— 098

5.3 掺杂二氧化钛单晶的电化学性能研究 —— 103

5.4 掺杂二氧化钛单晶的光催化性能研究 —— 104

第 6 章
低于相变温度自诱导合成两相比例可控的相结二氧化钛 —— 113

6.1 相结二氧化钛研究进展 —— 115

6.2 两相比例可控的相结二氧化钛的合成与表征 —— 116

6.3 相结二氧化钛纳米材料的光催化活性 —— 123

6.4 相结二氧化钛纳米材料的光催化活性机制研究 —— 126

第7章
硼掺杂诱导两相比例可控二氧化钛的光催化降解特性 —— 133

7.1 相结与掺杂共同作用的二氧化钛研究进展 —— 135

7.2 硼掺杂两相比例可控的二氧化钛的合成与表征 —— 136

7.3 硼掺杂两相二氧化钛纳米材料的光催化阿特拉津降解性能及降解路径分析 —— 141

7.4 硼掺杂两相二氧化钛纳米材料的光催化机制分析 —— 145

第8章
氧化锌/硫化镉/二氧化钛异质结光催化剂的构筑及其光催化降解特性 —— 151

8.1 二氧化钛与金属氧化物/硫化物异质结的研究进展 —— 153

8.2 氧化锌/硫化镉/二氧化钛异质结的合成与表征 —— 154

8.3 氧化锌/硫化镉/二氧化钛异质结的光催化阿特拉津降解性能及降解路径分析 —— 161

8.4 氧化锌/硫化镉/二氧化钛异质结的光催化机制分析 —— 170

第9章
二维二氧化钛/氮化碳异质结的构筑及其光电催化性能 —— 177

9.1 二氧化钛与非金属半导体异质结的研究进展 —— 179

9.2 二维二氧化钛与氮化碳复合材料的合成 —— 180

9.3 二维二氧化钛与氮化碳复合材料电荷密度散度的 DFT 计算 —— 181

9.4 二维二氧化钛与氮化碳复合材料的表征 —— 183

9.5 二维二氧化钛与氮化碳复合材料光电催化双酚 A 降解性能及降解路径分析 —— 190

9.6 二维二氧化钛与氮化碳复合材料的光电催化机制分析 —— 192

第 10 章
单颗粒荧光光谱和单分子成像技术解析二氧化钛光催化高活性位点 —— 199

10.1 二氧化钛催化机制的原位光谱研究进展 —— 201

10.2 高暴露晶面锐钛矿二氧化钛的合成与表征 —— 202

10.3 在宽场显微镜下单分子反应的荧光测试 —— 206

10.4 共聚焦显微镜下单颗粒荧光成像测定 —— 208

10.5 理论计算模拟高暴露晶面锐钛矿二氧化钛的活性位点 —— 210

第 1 章

绪 论

1.1 光催化的定义及二氧化钛材料的研究进展

在一定波长的入射光(紫外光、红外光或者可见光)的照射下,光催化剂吸收了入射光的能量之后改变了化学反应速率并引起了反应成分的化学改变,该过程一般被定义为光催化反应[1-2]。以光催化反应为基础的光催化技术已成为一种重要的高级氧化技术,在环境净化方面受到了广泛关注。与传统的生物降解法和其他物化处理技术如吸附法相比,光催化技术能够直接利用太阳光的能量提供强氧化基团,用来去除水中许多不可或难于生物降解的有机污染物,如染料、农药、杀虫剂等,并可以将它们矿化为 CO_2 和 H_2O,避免了二次污染[3]。

1972 年,日本东京大学 Fujishima 和 Honda 两位教授首次报道发现 N 型半导体 TiO_2 单晶电极光催化分解水从而产生氢气的现象,从而发现了 TiO_2 在光催化领域中的应用[4]。1976 年,Carey 等人就以 TiO_2 为光催化剂在紫外光(365 nm)照射下成功降解水中的多氯联苯(PCBs)[5]。后来 Anpo 等又发现其也可以降解氰化物离子[6]。此后 40 多年来,TiO_2 光催化剂纳米材料被公认为是环境治理和能源转化领域里具有较大潜力的材料[1]。TiO_2 的应用领域,从传统的颜料、化妆品、牙膏等转向新兴发展的光电化学电池[7]、染料敏化太阳能电池[8]、光催化剂[9]、光伏电池[10]、锂离子电池[9]、传感器[11]、生物医药处理[12]等。TiO_2 所有这些功能应用几乎关注 21 世纪人类所面临的三大最重要、最具有挑战性的问题——能源[7]、环境[13]和健康[12]。同时,基于 TiO_2 具有无毒、元素丰富、化学稳定性良好和易合成等明显优点,TiO_2 被世界范围内广泛研究,通过其物理化学性质实现各种功能。尤其涉及 TiO_2 在光催化和染料敏化太阳能电池的应用,极大地促进了 TiO_2 在可控物相、尺寸、形貌、缺陷和异质结方面的研究发展,如图 1.1 所示[14-18]。

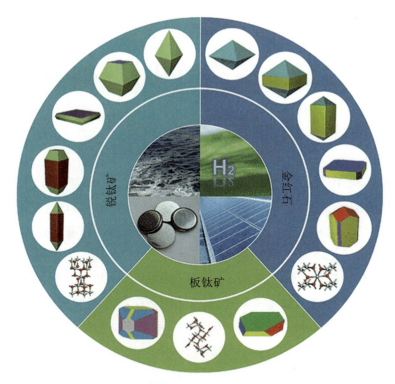

图1.1 TiO₂的结构与应用[16]

1.1.1
二氧化钛的晶体类型及催化机制

目前,已发现的 TiO₂ 有 11 个晶相,如图 1.2 所示[19]。其中,前面的 6 个晶相(从金红石到 TiO₂(R))是在室温或者低压下能稳定存在的。它们的密度范围从 ~3.5 g/cm³(TiO₂(H))到 ~4.2 g/cm³(金红石)。在图 1.2 中最后 5 个物相(TiO₂(Ⅱ)到立方体)是高压物相,密度范围从 ~4.3 g/cm³(TiO₂(Ⅱ))到 ~5.8 g/cm³(立方体)。所有的 TiO₂ 相都可以看作是由 Ti-O 多面体通过一个可变数量的共享角、边缘或者晶面连接建立起来的。低压下的 TiO₂ 多面体均呈正八面体,其中 TiO₂ 的 4 个相金红石、板钛矿、锐钛矿和 TiO₂(B)是在自然界中可以存在的物相[20-21]。对于金红石、板钛矿和锐钛矿来说,金红石是最具有热稳定性的 TiO₂,比锐钛矿和板钛矿的亚稳相具有更低的自由能[22];但是锐钛矿型的 TiO₂ 具有更高的导带能带和更低的电子空穴复合率,因此在光催化领域应用更为广泛[23]。同时，TiO_2纳米晶的物理化学性能不仅受固有电子结构的影响，而且还受其颗粒尺寸、形貌和表面特性等的影响[20,24]。

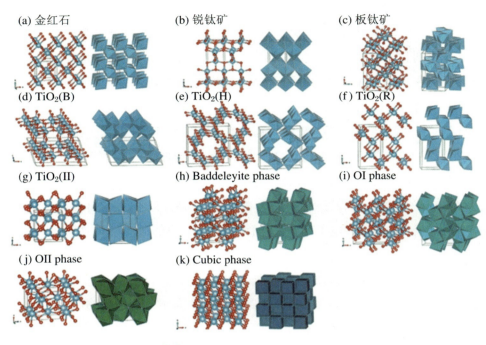

图 1.2　TiO_2晶体的不同物相[19]

1.1.2
二氧化钛光催化降解污染物的机制

在光催化剂吸收外加光源后，光催化剂价带上的电子即光生电子会被激发转移至导带，同时产生的光生空穴会被留在价带上，这时光催化剂表面就会产生所谓的光生电子（e^-）和光生空穴（h^+）[25]。这一阶段被称为半导体的"光激发态"，导带和价带之间的能量被称为"带隙"。带隙关系到半导体光催化剂能有效吸收的光的波长。以光解水为例，在光催化剂被激发后，电子和空穴发生分离并迁移至催化剂表面，光生电子作为还原剂，光生空穴作为氧化剂分别与水反应产生氢气和氧气。对于光催化降解污染物来说，光生电子转移到催化剂表面的溶解氧上形成超氧负离子，同时光生空穴会将吸附在催化剂表面的水氧化成为羟基自由基。因为超氧负离子和羟基自由基都具有强氧化性，所以理论上能将大部分有机污染物氧化至 CO_2 和 H_2O[25]。无机物半导体光催化的过程包括光吸收过程、载流子激发过程(电子和空穴)、载流子复合过程、载流子分离和迁移过程、载流子捕获和转移到水或其他分子上的过程。这些过程均影响了半导体光催化的最终效率[13]。

在光催化研究领域,可以将常见的半导体光催化剂大致分为紫外激发宽带隙光催化剂和窄带隙光催化剂两类[26]。由于半导体光催化剂的光吸收范围与其带隙存在 $K=1240/E_g$ 的关系,因此,宽带隙光催化剂主要是半导体吸收波长大部分在紫外区域的光催化剂。这一类的催化剂有很多,如 TiO_2、ZnO、$ZnWO_4$ 等。其中最典型的紫外激发宽带隙催化剂是金属氧族化合物 TiO_2。

TiO_2 的光催化反应机制如图 1.3 所示[26],当吸收的光能大于或等于 TiO_2 的带隙(E_g)时,价带上的电子被激发到空白的导带,留下了价带上的空穴(图中过程Ⅰ)。图中过程Ⅱ和过程Ⅲ是平行发生的,但是过程Ⅱ的复合比过程Ⅲ的迁移过程快很多。在光生电子和空穴转移到 TiO_2 表面的活性位点时则分别发生氧化/还原反应(过程Ⅳ和过程Ⅴ)。具体来说,在过程Ⅳ和Ⅴ中,在 TiO_2 降解污染物的反应中,此类反应为低能垒反应,反应的 $\Delta G<0$,价带上的空穴能跟水反应产生羟基自由基。羟基自由基是很强的氧化剂,能够实现污染物矿化。此外,空穴还能直接氧化有机污染物形成 R^+ [27-28]。对光解水来说,此反应为高能垒反应,反应的 $\Delta G>0$,水分子被光生电子还原形成氢气,同时光生空穴氧化水形成氧气。TiO_2 能够进行有效地光催化光解水的条件为:TiO_2 的导带底能级比 H^+/H_2(0 V vs. NHE)氧化还原电位更负,且价带顶能级比 O_2/H_2O(1.23 V vs. NHE)氧化还原电位更正。光催化解离水的最小带隙为 1.23 V,理论上所有满足这些条件的光催化剂都能成功发生光解水反应[29]。

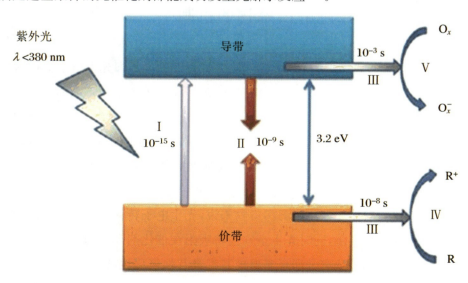

图 1.3　TiO_2 光催化反应机制[26]

由于 TiO_2 具有良好的光催化能力,其在多个领域的应用潜力被挖掘起来[30],如光伏、水分解、杀菌消毒、光合成、超亲水研究、环境净化等方面,其中关于环境净化的应用引起了人们的广泛关注和研究。

1.2 二氧化钛在环境催化领域中的应用

1.2.1 水体污染治理

水体污染主要来自日常生活、农业、养殖业及工业等所产生的废水。水体的污染物主要包括持久性有机污染物(POPs)、抗生素、杀虫剂、农药和染料等种类。尤其是工业废水中,大部分都含有有毒有害物质,会稳定持久残留于水体中,甚至会渗透至地下水体,造成地下水污染,给人们生活带来严重的安全威胁。由此,针对水体污染治理,纳米 TiO_2 的光催化技术,因具有的化学稳定性、高催化活性、成本低、无二次污染等优点,在环境污染的催化治理方面得到了广泛的应用,如图 1.4 所示。TiO_2 在光激发过程中,可以形成光生电子和空穴,电子可以与氧形成氧负离子,而空穴可以依靠其强氧化性与水反应形成羟基自由基,这样就可以对目标污染物进行矿化或选择性还原[31-34]。

已有文献报道,高暴露{001}晶面的 TiO_2 具有较强的光催化活性,能够在 UV 下快速降解硝基苯,并且有一定的矿化作用[35-36]。同样,TiO_2 作为一种光催化剂,在处理中性红、亚甲基蓝、活性黄等各种工业印染废水时,展示了较好的降解效果。另外针对农业污水,比如当前生产及使用量最大的有机磷所产生的废水毒性大、积累性较强,生物法降解难度大。TiO_2 薄膜材料的光催化技术,可以很好地回收利用,已被实际用于养殖废水的光催化高级氧化处理[37-38]。除此之外,针对石油行业的焦化废水和石油带来的油类污染物质等,也可以借助 TiO_2 光催化技术,利用太阳能来处理含油废水,经过长时间的处理,可以实现污水中的油质的高效去除。

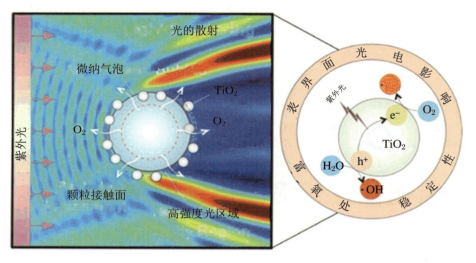

图 1.4　TiO_2 在水体污染治理方面的应用[31]

1.2.2 土壤污染治理

因化学工业呈规模化的生产发展,农药及化肥大量使用,工业污水、养殖业及种植业的污水等逐渐渗入土壤当中,致使土壤环境遭受严重的污染。而伴随着土壤污染问题逐渐恶化,会直接影响到土壤当中所有植物正常的生长,也会威胁到陆地上的生物特别是人类的生命安全[39]。TiO_2 光催化剂在水体及土壤中的污染控制方面表现出巨大的潜力(图 1.5),首先,因为这种高级氧化手段的能量来源是自然界中广泛存在的太阳光和氧气,且通常在常压下进行;其次,TiO_2 光催化在紫外光条件下具有非常优异的光催化活性;再次,因为 TiO_2 光催化剂在保证良好光催化能力的同时,还具有廉价易得、环境友好、良好的化学稳定性等优点;最后,在 TiO_2 光催化过程中无其他光生中间体化合物生成。在污染净化过程中,TiO_2 光催化纳米颗粒可以自由地在废水中沉降或者将其制备成薄膜[40-42]。经过大量试验研究发现,借助纳米 TiO_2 的光催化技术,能够提高土壤对污染物的自身吸附力,再通过搅拌并处于紫外线照射环境下,能够提升土壤及纳米颗粒自身受辐射的概率,从而使污染物的降解率相对较高。那么,土壤污染经过降解修复期间,除需利用可见光外,技术员需要尤为注意降解过程中间产物的有无毒性,以便于能够充分发挥纳米 TiO_2 的光催化科学技术优势,提升土壤污染整体的治理效果。

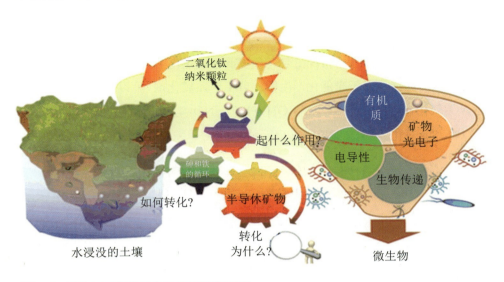

图 1.5　TiO_2 在环境土壤治理方面的应用[39]

1.2.3
气体污染治理

如图 1.6 所示，TiO_2 光催化剂目前已经被广泛用于环境气体治理[43]，如大量应用在室外建筑材料，例如油漆和铺路石，目的是降低空气中的废弃污染物，如二氧化碳（CO_2）、氮氧化物（NO_x）和有机挥发物等有害气体的浓度。众所周知，TiO_2 表面在紫外光照下会产生光生电子和光生空穴，然后生成强氧化性基团——羟基自由基和超氧自由基。早在 1979 年，Topalian 等人就报道了将 TiO_2 作为光催化剂还原水中 CO_2 制备甲酸、甲醛、甲醇和甲烷[44]。此后人们便常关注以 TiO_2 为催化剂还原 CO_2 的催化反应[6,45-47]。反应中生成的两种有机产物甲烷（CH_4）和甲醇（CH_3OH）是通过在 TiO_2 表面吸附的具有较高的电位（1.9 eV）的 CO_2 直接还原而成的。此外许多金属团簇，如 Cu、Pt 和 Pd 在 TiO_2 表面负载通常可以降低电子和空穴的复合，从而强化 CO_2 还原的效果[48-50]。在空气污染净化过程中，CO_2 可以转化为许多不同的碳氢化合物，例如，$CO_2 + H_2O \longrightarrow CH_4$、$CH_2O$、$CH_3COOH$ 等[46,48-52]。值得关注的是，吸附在 TiO_2 表面的 CO_2 和质子会产生电子迁移竞争[53]。最初电子传输的过程包含了 O＝C＝O 双键断裂以及氢原子的吸附，从而形成甲酸盐[46,52,54]。但同一时间的电子/质子传输过程导致了甲氧自由基的形成。最后生成的自由基可以通过单质子和双电子的反应进一步在 TiO_2 表面转化为 CH_4。在常压下,由于空气中 NO_x 的浓度很低,因此通过直接氧化法将 NO_x 转化为硝酸盐(NO_3^-)是一个非常缓慢的过程。但是加入 TiO_2 作为催化剂后,在空气环境下通过光化学氧化可以有效加速 NO_x 的氧化速度。这是由于光催化产生的·OH 是强氧化剂,可以直接在 TiO_2 表面把 NO_2 氧化为 NO_3^- [55]:$NO_2 + \cdot OH \rightarrow H^+ + NO_3^-$,另外光催化产生的·$O_2^-$ 也可以将 NO 氧化为 NO_3^-:$NO + \cdot O_2^- \rightarrow NO_3^-$。

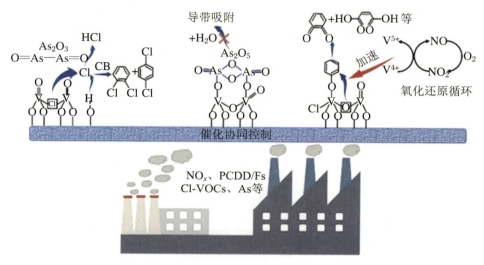

图 1.6 TiO_2 在环境气体治理方面的应用[43]

TiO_2 纳米光催化剂还可以用来降解有机挥发性物质[40],如 CH_3COCH_3 [56]、HCH_3OH [56]、HCHO[57] 和 CH_2Cl_2 [58]等,同样取得了良好效果,多数情况在紫外条件下 TiO_2 光催化反应的去除率可以达到 90% 以上。总之,TiO_2 显示出了良好的光催化处理污染物的巨大潜力。当前,各种产品形式纳米 TiO_2 的光催化剂被广泛应用在环境污染综合治理当中,如纳米的光催化剂类涂料、窗帘等,今后还会深入开发及研究出实际应用的一种光催化类产品。

1.2.4

二氧化钛光催化降解污染物存在问题

尽管光催化技术发展了许多年,光催化在水处理应用中面临的核心问题仍然是反应速率低。主要影响因素包括以下 4 个方面:① 溶液表面对入射光的反射;② 催化剂对入射光的吸收;③ 催化剂表面对入射光的反射;④ 光生电子的复合。另外还有一些其他因素,如溶液的光吸收问题、粉体催化剂的回收问题、废水的水质特性等。在以上4个主要方面中,溶液表面对入射光的反射问题可以通过光化学反应器的合理优化解决,催化剂表面对入射光的反射问题可以通过改变光催化剂表面的平面结构解决。因此,在光催化水处理中,人们研究的上游科学问题主要是催化剂对入射光的吸收以及光生电子的复合两大瓶颈。对于非改性的 TiO_2 光催化剂来说,TiO_2 作为光催化剂,主要存在3类问题:① 由于带隙的关系,TiO_2 光催化剂只能吸收紫外光,而紫外光只占太阳光的4%~5%,利用率有限,如图1.7所示[59];② TiO_2 表面的光生电子空穴容易复合,降低了有效的电荷分离[60-61];③ TiO_2 本身高活性位点的暴露有限,这也降低了其对水分子或污染物有效吸附催化的能力[62-63]。那么基于此,众多的学者开始研究 TiO_2 催化剂的改进[64-66]。一般只在紫外光下表现出较好的光催化活性,而自然光中只有5%左右的紫外光成分,极大地限制了 TiO_2 光催化在环境净化领域的实际应用。如何设计和合成性能优异、有实际应用潜力的 TiO_2 光催化剂,是 TiO_2 材料在环境净化领域发展和应用的一个重大挑战。

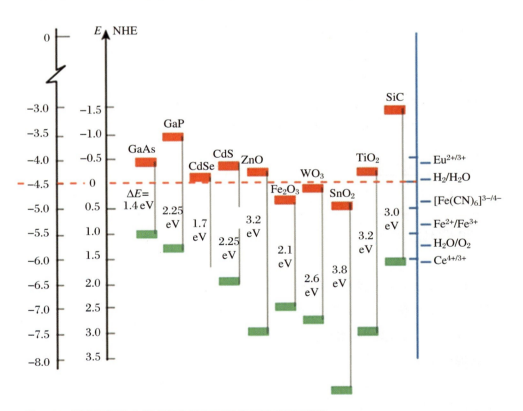

图1.7 常见半导体在标准氢电极下的导带与价带位置[59]

1.3 二氧化钛的晶体类型及催化机制

TiO$_2$光催化剂的活性通常受到光生电子空穴对快速复合的限制。最近有文献综述了光生载流子在光催化过程中复合的重要性[67]。尽管这个过程是不希望发生的，但是对它的研究可以提供关键的 TiO$_2$ 电荷载流子的动力学信息。

通常，由于存在载流子的各种可能性，光生载流子不会快速直接地复合，比如半导体表面存在空穴的捕获。实验中利用电子顺磁共振波谱（EPR）证明 TiO$_2$ 可以利用价带的空穴直接氧化表面的水形成羟基自由基[68-70]。同时，热力学分析也证明了这些实验结果可能发生的光反应[71]。因为氧化还原电位显示羟基自由基比 TiO$_2$ 价带边更正一点，所以原则上光生空穴可以产生这些活性物质。然而，Cheng 等人通过模型计算证明自由基在 TiO$_2$ 上的吸附势垒太高，认为自由基形成不太可能[72]。而随后的分子动力学模拟也证明了该结论[73]。然而，Imanashi 等人则从光电发射谱评估了对于 H$_2$O aq 和 OH aq—的 O 2p 轨道能力水平，从相应的紫外光电谱评估了对于 Ti—OH 和 TiO$_2$/H$_2$O 的界面的 O 2p 轨道能力水平[74-76]。基于以上的结果，光催化过程中，羟基自由基的形成不能完全被排除，这是由于光生电子可以和水中的溶解氧发生电还原，或在酸性溶液中由空穴捕获晶格连接的 O$_2^-$ 形成羟基自由基[77-80]。

1.3.1 晶面调控

已有文献报道很多方法可以用于合成清晰的 TiO$_2$ 晶面，如湿化学法（水热、溶剂热和非水解）、气体氧化法、拓扑置换法、无定型 TiO$_2$ 晶体转化法和外延生长法等[81-83]。在这之中，水热和溶剂热法由于其在晶体成核和生长的可控性，被广泛地用于合成特定暴露晶面。

TiO$_2$ 的许多应用迫切需要发展 TiO$_2$ 特定晶面的合成。尽管关于 TiO$_2$ 各种纳米结构的形状（纳米线[84]、纳米棒[83]、纳米带[85]、纳米管[86]、纳米球[87]和纳米片[88]）在过去几十年已经实现，但是对于清晰的高能晶面合成在 2008 年前比较少,如图 1.8 所示[35,81]。实验发现,在 TiO$_2$ 形成的过程中,F$^-$ 粒子会吸附在晶体表面,降低了形成 TiO$_2${001} 晶面的表面能垒,从而使该晶面高暴露显现。并且,进一步的研究证明,随着反应时间的增加,TiO$_2${001} 晶面的也会被 F$^-$ 粒子刻蚀,留下{101}晶面[89]。

图 1.8 高暴露{001}晶面的 TiO$_2$ 的 SEM 图[81]

1.3.2
元素掺杂

阳离子掺杂作为一种常规的方法被应用于半导体的改性。在 TiO$_2$ 的能带结构中,O 2p 轨道占据价带(VB),而 Ti 3d、4s、4p 轨道在未占用的导带(CB)。CB 的较低位置由 Ti 3d 轨道决定。一旦阳离子取代 Ti,杂质水平可以被引入禁带。这种中间能量水平可以作为电子受体或供体,使 TiO$_2$ 吸收可见光。直到现在,为了制备可见光响应的 TiO$_2$,很多人研究了阳离子掺杂的方法[90]。

同样,阴离子掺杂可以抬升 TiO$_2$ 的能带结构中价带的位置,减小相应的禁带宽度,使 TiO$_2$ 具有可见光活性[91-92]。据文献报道,可以通过调控红色的 TiO$_2$ 中 B 从体相掺杂到表面掺杂,实现可见光光催化活性[93]。同时,高温下,H 的掺杂更是大大提升了可见光的光催化效率[94-95]。

1.3.3 异质结的构筑

异质结是两种不同的半导体相接触所形成的界面区域。按照两种材料的导电类型不同,异质结可分为同型异质结(p-p结或n-n结)和异型异质(p-n)结,多层异质结称为异质结构。异质结常具有两种半导体各自的p-n结都不能达到的优良的光电特性。而半导体异质结构有以下几个方面的基本特性:① 量子效应。电子的特性会受到量子效应的影响而改变。② 迁移率变大。在异质结构中,可将杂质加在两边的夹层中,该杂质所贡献的电子会掉到中间层,因其有较低的能量,所以电子的行动就不会因杂质的碰撞而受到限制,因此其迁移率就可以大大增加。③ 奇异的二度空间特性。因为电子只剩下两个自由度的空间,半导体异质结构因而提供了一个非常好的物理系统可用于研究低维度的物理特性[96-97]。

而以 TiO_2 材料为基础的异质结复合材料也是一种改进 TiO_2 光催化活性的一种方法,如图1.9所示。通过文献调研发现,以金属、金属氧化物(硫化物、磷化物)、碳材料作为 TiO_2 复合材料有广泛的研究和应用[96-97]。基于 TiO_2 的带隙宽和电荷能力分离差,复合材料一方面可以提高电荷的分离能力,另一方面也可

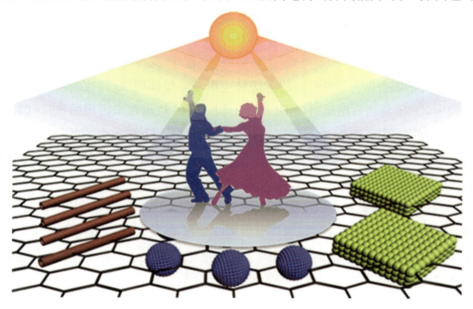

图1.9　材料复合的示意图[96]

以提高可见光活性。金属可以作为有效的助催化剂，提高电子空穴的分离能力[98]。而金和银不但可以作为助催化剂，同时也可以利用等离子共振吸收可见光，导出热电子转移至 TiO_2，使其拥有可见光活性[99-101]。另外，CdS 或 ZnS 作为典型的可见光催化剂，可以通过与 TiO_2 的复合提高 TiO_2 的可见光活性[102-104]。而 NiS、CoS 等可以作为 TiO_2 的助催化剂，提高电子空穴分离能力，有利于产氢产氧[105-107]。同样，近几年过渡金属磷化物作为优异的产氢产氧催化剂，与 TiO_2 复合，也能提高 TiO_2 电子空穴分离能力[108]。

碳材料拥有高稳定、良好的导电性和高比表面积等优点，在光催化过程中，不仅是良好的光催化剂载体，而且还是良好的导电材料，可以有效将光生电子导出。而 C_{60}、碳纳米管、石墨烯作为比较新兴的碳材料，被广泛地用于复合材料的研究[109-113]。已有大量文献报道，石墨烯与 TiO_2 的复合材料，不仅可以提高 TiO_2 的分散性，而且可以增强其对光的吸收；同时瞬态光谱上研究发现石墨烯可以提高光生载流子的寿命，从而提高光催化活性。而石墨烯与 CdS 的复合体系，提高了材料的循环稳定性，说明了石墨烯增强了 CdS 的抗光蚀作用[114-115]。

1.3.4
分子印迹的构筑

分子印迹，又称"分子烙印"(molecular imprinting)，由 20 世纪 30 年代 Breinl、Haurowitz 和 Mudd 提出的抗体理论和 20 世纪 40 年代 Pauling 对上述理论的进一步阐述而奠定基础。随着 1949 年 Dickey 首先提出"专一性吸附"概念和 1972 年 Wulff 研究小组报道的人工合成分子印迹聚合物，这项技术才渐渐被人们所熟悉。[116]。近十多年来，分子印迹技术逐渐被人们广泛地应用于临床诊断、色谱分析以及传感器等领域[117]。分子印迹材料像真实抗体一样具有特异性识别功能，因此被称为"模拟抗体"[118-119]。与传统的光催化选择性改善技术相比，分子印迹具有选择性高、适用范围广和针对性强的优点。通过在光催化剂上构筑具有识别位点的修饰层，可以实现特定化合物的光电化学识别和选择性催化去除。

有研究报道称，以 2-硝基苯酚为模板合成的复合催化材料 $Fe_3O_4/Al_2O_3/$2-NP-TiO_2，通过对比实验和干扰实验，证明其对 2-硝基苯酚具有高效的选择性降解[120]。同样，对于低溶解度高毒性的污染物，以结构类似物作为分子印迹模板合成具有分子印迹的 TiO_2[121]，以硝基苯为例，成功实现了选择性降解硝基苯的目标。近些年，又有连续的研究报道称，通过液相沉积的方法合成了分子印迹型光催化剂,如以雌酮为模板的 TiO_2/Fe_3O_4 复合催化剂,以盐酸四环素为模板修饰的印迹 TiO_2 膜,以水杨酸为模板的分子印迹型 TiO_2 光催化薄膜等[122-124],分别实现了对目标污染物的有效光催化降解。进一步研究发现,通过分子印迹技术,壳聚糖的生物吸附与 TiO_2 的光催化技术结合起来,以甲基橙为模板成功合成了一种新的有机-无机复合材料,即表面分子印迹壳聚糖-TiO_2[125]。研究数据表明该电极对甲基橙的选择性光催化降解能力增强,并且该催化剂可以被重复利用,体现了分子印迹的稳定性。

在此基础上,如图 1.10 所示,通过分子印迹技术合成的光电催化材料,成功地被用于光电催化污染物的选择性降解[116,126-127]。研究人员从 2,4-二氯苯氧乙酸分子印迹的一维 SnO_2 纳米棒[128],发展到该分子印迹的二维高暴露{001}晶面 TiO_2[129],通过暴露更多的分子印迹活性位点,提高了对环境污染物 2,4-二氯苯氧乙酸的光电催化效率。

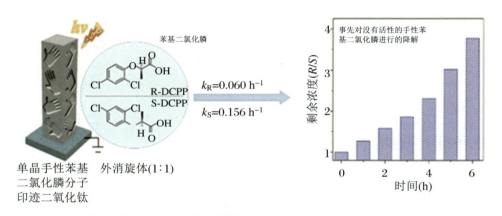

图 1.10　手性分子印迹 TiO_2 光电催化[116]

1.4 二氧化钛在催化过程中的原位/微观研究

TiO_2 的原位/微观成像研究在最近几年也有飞速发展[130-133]。扫描隧道显微镜(STM)被广泛地用来研究 H_2O 分子等与 TiO_2 表面的微观光催化反应机制[134]。同样,光学显微镜也更多地被用来研究 TiO_2 光催化反应活性及其动力学。并且提高光学显微镜的分辨率,可以在更符合实际的体系下观测在 TiO_2 表面的分子反应、荧光光谱与荧光寿命,从而得到实际的光催化活性位点[135]。

1.4.1
荧光显微镜单分子成像研究二氧化钛表面催化反应动力学

在光催化过程中,有效的光生电子空穴转移对光催化性能有重要的意义。那么,在光催化体系中,原位观测光生电子的转移就可以深入了解电子在半导体表面的迁移和复合过程[133]。TiO_2 作为典型的半导体光催化剂,通过晶面间的带隙位置,默认{001}为氧化晶面,而{101}为还原晶面,光生电子从{001}晶面转移至{101}晶面。并且已有文献报道利用单分子荧光实验证明了光生电子确实可以发生在晶面间的转移,从而直观地证明该结论的正确性,如图 1.11 所示[135]。

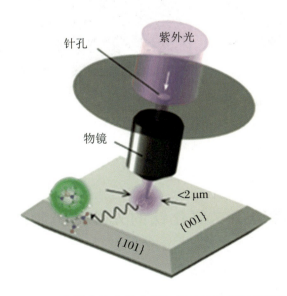

图 1.11 染料分子在单个 TiO_2 表面的反应显微镜观察示意图[135]

1.4.2 共聚焦显微镜荧光寿命成像技术研究单个颗粒表面荧光寿命分布

近年来,对于单一颗粒的原位荧光表征技术也被迅速发展并应用于材料化学。在光催化过程中,均相的反应只能表现为整体的光催化性能。对于单一颗粒进行的原位表征可以更好地研究催化剂的本征催化性能。已有文献报道,利用该相关单粒子光谱,以研究 Pt-Au NRs 的光催化水分解过程为例,通过单颗粒荧光光谱分析,表明电子从 Au NRs 转移到 Pt[131]。同理,当 Pd 取代 Pt 时,如图 1.12 所示,采用高灵敏度的单粒子荧光光谱同样得到了相关的电子转移路径[132]。并且该技术也被广泛地用于纳米材料的生物成像,如生物标记、单颗粒追踪等方面[136-138]。

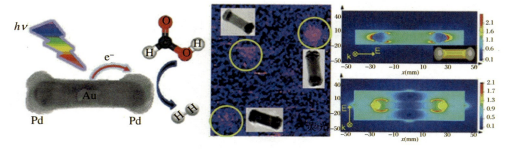

图 1.12　单个含 Pd 负载的 Au NRs 的荧光成像图[132]

1.4.3 扫描隧道显微镜中二氧化钛的光催化机制研究

由于吸附作用在 TiO_2 材料表面化学的过程有重要作用,那么通过原位的扫描隧道显微镜观察 TiO_2 表面吸附的 O_2、CO、H_2O 等分子在光催化过程中发生的变化将对光催化反应机制研究有重大意义,如图 1.13 所示[139-141]。已有文献报道,通过扫描隧道显微镜可以观察到在桥连的氧空位或在 $TiO_2\{110\}$ 晶面的羟基表面上[139]有吸附的氧分子。实验结果提供了直接的证据说明氧分子能固定地吸附在氧空位和羟基上。同时在 TiO_2 表面覆盖的氧分子减少时,吸附在羟基上的氧分子比吸附在氧空位上的更加稳定。而且氧空位上覆盖的氧分子越多,吸附的氧分子越稳定。而基于有应用前景的水分解反应光催化剂 TiO_2,直接地观察光催化水分子在 TiO_2 表面的分解过程则为深入研究半导体光催化产氢的机制提供了真实的依据,为设计合理的催化剂提供了充足的实验支持[140]。但是,反应顺序一直存在令人困惑的问题,如光催化最初的反应步骤是什么? 如何发生? 据文献报道,低温扫描隧道显微镜,可以原位观测单独吸附在 TiO_2 {110}晶面的水分子在 UV 照射下的分解反应。同时,这个实验清楚地解释了光催化过程中所产生的 2 种自由基,并且实验过程也清楚呈现了反应在 TiO_2 表面发生的具体位置[141]。

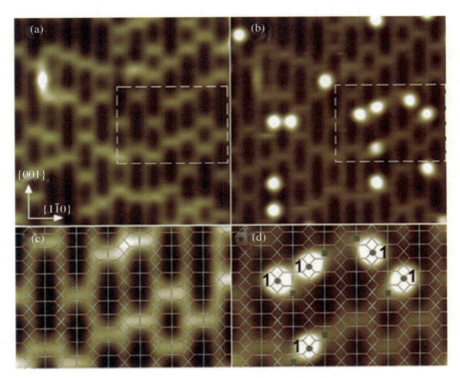

图 1.13　STM 下观测的 CO 吸附在金红石{110}晶面上的照片[139]

基于以上研究,对近几十年的 TiO_2 的不同表面的化学反应性已被广泛研究[141]。通过 STM 同样可以研究得到相应的表面活性位点。实验证明,锐钛矿{001}表面的反应活性与其还原程度有关。

1.4.4
表面光电压光谱技术利用颗粒表面的电压变化研究催化活性机制

当光生载流子在空间上分开时,某些电荷转移或重新分布而产生光电压,从而导致样品表面的接触电势差(CPD)发生变化,因此称该光电压为表面光电压(SPV)。SPV 严格地定义为光照引起的表面电势变化。从理论上讲,表面光伏效应与半导体空间电荷区(SCR)中的电荷分离和传输有关。最近有报道称,可以通过使用空间分辨 SPV 显微镜直接测量光催化剂表面和内部表面上的光生电荷载体[142]。在光催化中,光生的电荷分离和转移作为能量泵传输的关键步骤,决定了整体太阳能转换效率。SPV 技术也属于研究光催化中光生的电荷分离和转移过程的先进技术。与其他技术相比,SPV 信号直接与光生载流子在空间中的分离,包括光生电荷载体表面的演变过程[143-145]有关。

此外,如图 1.14 所示,基于空间分辨 SPV 技术的开尔文探针力显微镜(KPFM)可应用于纳米材料的电荷分离成像,研究单一材料及复合材料之间的电子空穴分离[145-148]。同时,通过原子力显微镜测量光电极的电流映射图像来反应材料的光导电性,从侧面证明了材料的电子传输能力[146,149-151]。

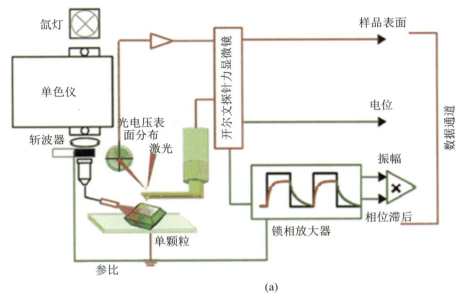

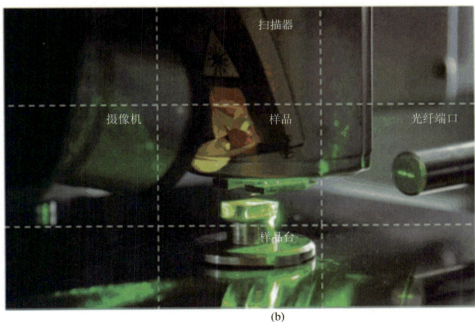

图 1.14 关键组件的示意图:(a) 照明系统、KPFM 系统和用于数据采集的数据采集系统;(b) 空间分辨 SPV 显微镜局部结构的照片[146]

参考文献

[1] CHEN X, SELLONI A. Introduction: titanium dioxide (TiO$_2$) nanomaterials[J]. Chem Rev., 2014, 114:9281-9282.

[2] 蔡伟民,龙明策.环境光催化材料与光催化净化技术[M].上海:上海交通大学出版社,2010.

[3] LONG Z, LI Q, WEI T. Historical development and prospects of photocatalysts for pollutant removal in water[J]. J. Hazard. Mater., 2020, 395:122599.

[4] FUJISHIMA A, HONDA K. Electrochemical photolysis of water at a semiconductor electrode[J]. Nature., 1972, 238:37-38.

[5] CAREY J H, LAWRENCE J, TOSINE H M. Photodechlorination of PCB's in the presence of titanium dioxide in aqueous suspension[J]. Bull. Environ. Contam. Toxical, 1976, 16: 697-701.

[6] ANPO M, YAMASHITA H, ICHIHASHI Y. Photocatalytic reduction of CO_2 with H_2O on titanium oxides anchored within micropores of zeolites: effects of the structure of the active sites and the addition of Pt[J]. J. Phys. Chem. B., 1997, 101:2632-2636.

[7] YI M, WANG X, JIA Y L. Titanium dioxide-based nanomaterials for photocatalytic fuel generations[J]. Chem Rev., 2014, 114:9987-10043.

[8] GRäTZEL M. Recent advances in sensitized mesoscopic solar cells[J]. Acc. Chem. Res., 2009, 42:1788-1798.

[9] ASAHI R, MORIKAWA T, IRIE H. Nitrogen-doped titanium dioxide as visible-light-sensitive photocatalyst: designs, developments, and prospects [J]. Chem Rev., 2014, 114:9824-9852.

[10] BAI Y, MORA-SERó I, ANGELIS F D. Titanium dioxide nanomaterials for photovoltaic applications[J]. Chem Rev., 2014, 114:10095-10130.

[11] BAI J, ZHOY B X. Titanium dioxide nanomaterials for sensor applications [J]. Chem Rev., 2014, 114:10131-10176.

[12] RAJH T, DIMITRIJEVIC N M, BISSONNETTE M. Titanium dioxide in the service of the biomedical revolution[J]. Chem Rev., 2014, 114: 10177-10216.

[13] HOFFMANN M R, MARTIN S T, CHOI W. Environmental applications of

semiconductor photocatalysis[J]. Chem Rev., 1995, 95:69-96.

[14] WANG X, LI Z, SHI J. One-dimensional titanium dioxide nanomaterials: nanowires, nanorods, and nanobelts[J]. Chem Rev., 2014, 114: 9346-9384.

[15] LEE K, MAZARE A, SCHMUKI P. One-dimensional titanium dioxide nanomaterials: nanotubes[J]. Chem Rev., 2014, 114:9385-9454.

[16] LIU G, YANG H G, JIAN P. Titanium dioxide crystals with tailored facets [J]. Chem Rev., 2014, 114:9559-9612.

[17] FATTAKHOVA-ROHLFING D, ZALESKA A, BEIN T. Three-dimensional titanium dioxide nanomaterials[J]. Chem Rev., 2014, 114:9487-9558.

[18] ZHANG N, YANG M Q, LIU S. Waltzing with the versatile platform of graphene to synthesize composite photocatalysts[J]. Chem Rev., 2015, 115:10307-10377.

[19] ZHANG H, BANFIELD J F. Structural characteristics and mechanical and thermodynamic properties of nanocrystalline TiO_2[J]. Chem Rev., 2014, 114:9613-9644.

[20] BLNRRNR J, VBNR D. Conversionof perovskiteto anataseand TiO_2(B): a TEM study and the use of fundamental building blocks for understanding relationship samong the TiO_2 minerals[J]. Am. Mineral., 1992, 77: 545-557.

[21] BANFIELD J F, VEBLEN D R, SMITH D J. The identification of naturally occurring TiO_2(B) by structure determination using high-resolution electron microscopy, image simulation, and distance-least-squares refinement[J]. Am. Mineral., 1991, 76:343-353.

[22] MLLER U. Inorganic Structural Chemistry[M]. New Jersey: Wiley, 2006.

[23] MATSUI M, AKAOGI M. Molecular dynamics simulation of the structural and physical properties of the four polymorphs of TiO_2[J]. Mol Simul., 1991, 6:239-244.

[24] BANFIELD J F. Oriented attachment and growth, twinning, polytypism, and formation of metastable phases: insights from nanocrystalline TiO_2[J]. Am. Mineral., 1998, 83:1077-1082.

[25] MILLS A, DAVIES R H, WORSLEY D. Water purification by semiconductor photocatalysis[J]. Chem. Soc. Rev, 1993, 22:417-425.

[26] WANG Y, WANG Q, ZHAN X. Visible light driven type Ⅱ heterostruc-

tures and their enhanced photocatalysis properties: a review [J]. Nanoscale., 2013, 5:8326-8339.

[27] CHEN D, CHENG Y, ZHOU N. Photocatalytic degradation of organic pollutants using TiO_2-based photocatalysts: a review[J]. J. Clean. Prod., 2020, 268:121725.

[28] GOPINATH K P, MADHAV N V, KRISHNAN A. Present applications of titanium dioxide for the photocatalytic removal of pollutants from water: a review[J]. J. Environ. Manage., 2020, 270:110906.

[29] WALTER M G, WARREN E L, MCKONE J R. Solar water splitting cells [J]. Chem Rev., 2010, 110:6446-6473.

[30] LAN Y, LU Y, REN Z. Mini review on photocatalysis of titanium dioxide nanoparticles and their solar applications[J]. Nano Energy., 2013, 2:1031-1045.

[31] FAN W, LI Y, WANG C. Enhanced photocatalytic water decontamination by micro-nano bubbles: measurements and mechanisms[J]. Environ. Sci. Technol., 2021, 55(10):7025-7033.

[32] BONANNI S, AIT-MANSOUR K, HARBICH W. Effect of the TiO_2 reduction state on the catalytic CO oxidation on deposited size-selected Pt clusters[J]. J. Am. Chem. Soc., 2012, 134:3445-3450.

[33] MOHAMED O S. Photocatalytic oxidation of selected fluorenols on TiO_2 semiconductor[J]. J. Photochem. Photobiol. A., 2002, 152:229-232.

[34] IZUMI Y, KONISHI K, MIYAJIMA T. Photo-oxidation over mesoporous V-TiO_2 catalyst under visible light monitored by vanadium $K\beta_{5,2}$-selecting XANES spectroscopy[J]. Mater. Lett., 2008, 62(617):861-864.

[35] YANG H G, SUN C H, QIAO S Z. Anatase TiO_2 single crystals with a large percentage of reactive facets[J]. Nature., 2008, 453:638-641.

[36] PAN J, LIU G, LU G Q. On the true photoreactivity order of {001}, {010}, and {101} facets of anatase TiO_2 crystals[J]. Angew. Chem. Int. Ed., 2011, 50:2133-2137.

[37] ZHANG A Y, WANG W K, PEI D Nl. Degradation of refractory pollutants under solar light irradiation by a robust and self-protected ZnO/CdS/TiO_2 hybrid photocatalyst[J]. Water Res., 2016, 92:78-86.

[38] ZHANG, YONG A, LONG. Electrochemical degradation of refractory pollutants using TiO_2 single crystals exposed by high-energy {001} facets[J].

Water Res., 2014, 66:273-282.

[39] CHEN Z, LIU Y, ZHANG C. Titanium dioxide nanoparticles induced an enhanced and intimately coupled photoelectrochemical-microbial reductive dissolution of As(Ⅴ) and Fe(Ⅲ) in flooded arsenic-enriched soils[J]. ACS Sustain. Chem. Eng., 2019, 7:13236-13246.

[40] DALTON J S, JANES P A, JONES N G. Photocatalytic oxidation of NO_x gases using TiO_2: a surface spectroscopic approach[J]. Environ. Pollut., 2002, 120:415-422.

[41] ALBERICI R M, JARDIM W F. Photocatalytic destruction of VOCS in the gas-phase using titanium dioxide[J]. Appl. Catal. B., 1997, 14:55-68.

[42] RAO, GOPALA G. Newer aspects of nitrification: I[J]. Soil Sci., 1934, 38(2):143-159.

[43] LONG Y, SU Y, XUE Y, et al. V_2O_5-WO_3/TiO_2 Catalyst for efficient synergistic control of NO_x and chlorinated organics: insights into the arsenic effect[J]. Environ. Sci. Technol., 2021, 55(13). DOI:10.1021/acs.ecf.1c02636.

[44] TOPALIAN Z, NIKLASSON G A, GRANQVIST C G, et al. Photo-fixation of SO_2 in nanocrystalline TiO_2 films prepared by reactive DC magnetron sputtering[J]. Thin Solid Films, 2009, 518:1341-1344.

[45] MENG A, CHENG B, TAN H, et al. TiO_2/polydopamine S-scheme heterojunction photocatalyst with enhanced CO_2-reduction selectivity[J]. Appl. Catal. B., 2021, 289:120039.

[46] WU J C S, LIN H M, LAI C L. Photo reduction of CO_2 to methanol using optical-fiber photoreactor[J]. APPL CATAL A-GEN., 2005, 296:194-200.

[47] WANG Z W, WAN Q, SHI Y Z, et al. Selective photocatalytic reduction CO_2 to CH_4 on ultrathin TiO_2 nanosheet via coordination activation[J]. Appl. Catal. B., 2021, 288:120000.

[48] WOOLERTON T W, SHEARD S, REISNER E, et al. Efficient and clean photoreduction of CO_2 to CO by enzyme-modified TiO_2 nanoparticles using visible light[J]. J. Am. Chem. Soc., 2010, 132:2132-2133.

[49] WANG A, WU S, DONG J, et al. Interfacial facet engineering on the Schottky barrier between plasmonic Au and TiO_2 in boosting the photocatalytic CO_2 reduction under ultraviolet and visible light irradiation[J]. Chem. Eng. J., 2021, 404:127145.

[50] MORADI M, KHORASHEH F, LARIMI A. Pt nanoparticles decorated Bi-doped TiO_2 as an efficient photocatalyst for CO_2 photo-reduction into CH_4 [J]. Sol. Energy, 2020, 211:100-110.

[51] YUI T, KAN A, SAITOH C, et al. Photochemical reduction of CO_2 using TiO_2: effects of organic adsorbates on TiO_2 and deposition of Pd onto TiO_2 [J]. ACS Appl. Mater. Interfaces., 2011, 3:2594-2600.

[52] DIMITRIJEVIC N M, VIJAYAN B K, POLUEKTOV O G, et al. Role of water and carbonates in photocatalytic transformation of CO_2 to CH_4 on titania[J]. J. Am. Chem. Soc., 2011, 133:3964-3971.

[53] NGUYEN T V, WU J C S, CHIOU C H. Photoreduction of CO_2 over ruthenium dye-sensitized TiO_2-based catalysts under concentrated natural sunlight[J]. Catal. Commun., 2008, 9:2073-2076.

[54] LINSEBIGLER A L, LU G, YATES JR J T. Photocatalysis on TiO_2 surfaces: principles, mechanisms, and selected results[J]. Chem. Rev., 1995, 95:735-758.

[55] KOČÍ K, OBALOVÁL, LACNÝ Z. Photocatalytic reduction of CO_2 over TiO_2 based catalysts[J]. Chem. Pap., 2008, 6:1-9.

[56] HO C-C, KANG F, CHANG G-M, et al. Application of recycled lanthanum-doped TiO_2 immobilized on commercial air filter for visible-light photocatalytic degradation of acetone and NO[J]. Appl. Surf. Sci., 2019, 465: 31-40.

[57] TOPALIAN Z, NIKLASSON G A, GRANQVIST C G, et al. Photo-fixation of SO_2 in nanocrystalline TiO_2 films prepared by reactive DC magnetron sputtering[J]. Thin Solid Films, 2009, 518:1341-1344.

[58] JOSSET S, TARANTO J, KELLER N, et al. UV-A photocatalytic treatment of high flow rate air contaminated with Legionella pneumophila[J]. Catal. Today., 2007, 129:215-222.

[59] Gratzel M. Photoelectrochemical Cells[J]. Nature, 2001, 414: 338-344.

[60] LI R, ZHANG F, WANG D, et al. Spatial separation of photogenerated electrons and holes among {010} and {110} crystal facets of $BiVO_4$[J]. Nat. Commun., 2013, 4:1432.

[61] CHEN X, SHEN S, GUO L, et al. Semiconductor-based photocatalytic hydrogen generation[J]. Chem. Rev., 2010, 110:6503-6570.

[62] NOWOTNY J, BAK T, NOWOTNY M K, et al. TiO_2 surface active sites

for water splitting[J]. J. Phys. Chem. B., 2006, 110:18492-18495.

[63] FUJITANI T, NAKAMURA I. Mechanism and active sites of CO oxidation over single-crystal Au surfaces and a Au/TiO$_2$ {110} model surface[J]. Chinese J. Catal., 2016, 37:1676-1683.

[64] NIU S, HUYAN H, LIU Y, et al. Bandgap control via structural and chemical tuning of transition metal perovskite chalcogenides[J]. Adv. Mater., 2017, 29: 1604733.

[65] NECHACHE R, HARNAGEA C, LI S, et al. Bandgap tuning of multiferroic oxide solar cells[J]. Nat. Photonics. 2014, 9:61-67.

[66] ZHOU T, ANDERSON R T, LI H, et al. Bandgap tuning of silicon quantum dots by surface functionalization with conjugated organic groups[J]. Nano Lett., 2015, 15:3657-3663.

[67] ANGELIS F D, VALENTIN C D, FANTACCI S, et al. Theoretical studies on anatase and less common TiO$_2$ phases: bulk, surfaces, and nanomaterials[J]. Chem. Rev., 2014, 114:9708-9753.

[68] RAJH T, OSTAFIN A E, MICIC O I, et al. Surface modification of small particle TiO$_2$ colloids with cysteine for enhanced photochemical reduction: an EPR study[J]. J. Phys. Chem., 1996, 100:4538-4545.

[69] CORONADO J M, MAIRA A J, CONESA J C, et al. EPR study of the surface characteristics of nanostructured TiO$_2$ under UV irradiation[J]. Langmuir, 2001, 17:5368-5374.

[70] ANPO M, SHIMA T, KUBOKAWA Y. ESR and photoluminescence evidence for the photocatalytic formation of hydroxyl radicals on small TiO$_2$ particles[J]. Chem. Lett., 1985, 14:1799-1802.

[71] WARDMAN P. Reduction potentials of one-electron couples involving free radicals in aqueous solution[J]. J. Phys. Chem. Ref. Data., 1989, 18: 1637-1755.

[72] CHENG J, VANDEVONDELE J, SPRIK M. Identifying trapped electronic holes at the aqueous TiO$_2$ interface[J]. J. Phys. Chem. C., 2014, 118: 5437-5444.

[73] DIESEN V, JONSSON M. Formation of H$_2$O$_2$ in TiO$_2$ photocatalysis of oxygenated and deoxygenated aqueous systems: a probe for photocatalytically produced hydroxyl radicals[J]. J. Phys. Chem. C, 2014, 118:10083-10087.

[74] IMANISHI A, OKAMURA T, OHASHI N, et al. Mechanism of water

photooxidation reaction at atomically flat TiO$_2$(rutile) {110} and {100} surfaces: dependence on solution pH[J]. J. Am. Chem. Soc., 2007, 129: 11569-11578.

[75] OTTOSSON N, FAUBEL M, BRADFORTH S E, et al. Photoelectron spectroscopy of liquid water and aqueous solution: electron effective attenuation lengths and emission-angle anisotropy[J]. J. Electron Spectros. Relat. Phenomena., 2010, 177:60-70.

[76] NAKAMURA R, NAKATO Y. Primary intermediates of oxygen photoevolution reaction on TiO$_2$(rutile) particles, revealed by in situ FTIR absorption and photoluminescence measurements[J]. J. Am. Chem. Soc., 2004, 126: 1290-1298.

[77] BERGER T, STERRER M, DIWALD O, et al. Light-induced charge separation in anatase TiO$_2$ particles[J]. J. Phys. Chem. B., 2005, 109: 6061-6068.

[78] MICIC O I, ZHANG Y, CROMACK K R, et al. Trapped holes on TiO$_2$ colloids studied by electron paramagnetic resonance[J]. J. Phys. Chem., 1993, 97:7277-7283.

[79] ISHIBASHI K-I, FUJISHIMA A, WATANABE T, et al. Quantum yields of active oxidative species formed on TiO$_2$ photocatalyst[J]. J. Photoch. Photobio. A., 2000, 134:139-142.

[80] HOWE R F, GRATZEL M. EPR study of hydrated anatase under UV irradiation[J]. J. Phys. Chem., 1987, 91:3906-3909.

[81] YANG H G, SUN C H, QIAO S Z, et al. Anatase TiO$_2$ single crystals with a large percentage of reactive facets[J]. Nature, 2008, 453:638.

[82] ANDERSSON M, ÖSTERLUND L, LJUNGSTRÖM S, et al. Preparation of nanosize anatase and rutile TiO$_2$ by hydrothermal treatment of microemulsions and their activity for photocatalytic wet oxidation of phenol[J]. J. Phys. Chem. B, 2002, 106:10674-10679.

[83] LIU B, AYDIL E S. Growth of oriented single-crystalline rutile TiO$_2$ nanorods on transparent conducting substrates for dye-sensitized solar cells [J]. J. Am. Chem. Soc., 2009, 131:3985-3990.

[84] WANG G, WANG H, LING Y, et al. Hydrogen-treated TiO$_2$ nanowire arrays for photoelectrochemical water splitting[J]. Nano Lett., 2011, 11: 3026-3033.

[85] WANG J, TAFEN D N, LEWIS J P, et al. Origin of photocatalytic activity of nitrogen-doped TiO$_2$ nanobelts[J]. J. Am. Chem. Soc., 2009, 131: 12290-12297.

[86] PARK J H, KIM S, BARD A J. Novel carbon-doped TiO$_2$ nanotube arrays with high aspect ratios for efficient solar water splitting[J]. Nano Lett., 2006, 6:24-28.

[87] LI N, LIU G, ZHEN C, et al. Battery performance and photocatalytic activity of mesoporous anatase TiO$_2$ nanospheres/graphene composites by template-free self-assembly[J]. Adv. Funct. Mater., 2011, 21:1717-1722.

[88] GUO W, ZHANG F, LIN C, et al. Direct growth of TiO$_2$ nanosheet arrays on carbon fibers for highly efficient photocatalytic degradation of methyl orange[J]. Adv. Mater., 2012, 24:4761-4764.

[89] YANG H G, LIU G, QIAO S Z, et al. Solvothermal synthesis and photoreactivity of anatase TiO$_2$ nanosheets with dominant {001} Facets[J]. J. Am. Chem. Soc., 2009, 131:4078-4083.

[90] MATSUMOTO Y, MURAKAMI M, SHONO T, et al. Room-temperature ferromagnetism in transparent transition metal-doped titanium dioxide[J]. Science, 2001, 291:854-856.

[91] DI VALENTIN C, PACCHIONI G, SELLONI A, et al. Characterization of paramagnetic species in n-doped TiO$_2$ powders by EPR spectroscopy and DFT calculations[J]. J. Phys. Chem. B, 2005, 109:11414-11419.

[92] ASAHI R, MORIKAWA T, OHWAKI T, et al. Visible-light photocatalysis in nitrogen-doped titanium oxides[J]. Science, 2001, 293:269-271.

[93] LIU G, YIN L-C, WANG J, et al. A red anatase TiO$_2$ photocatalyst for solar energy conversion[J]. Energy Environ. Sci., 2012, 5:9603-9610.

[94] ZHOU W, LI W, WANG J-Q, et al. Ordered mesoporous black TiO$_2$ as highly efficient hydrogen evolution photocatalyst[J]. J. Am. Chem. Soc., 2014, 136:9280-9283.

[95] CHEN X, LIU L, YU P Y, et al. Increasing solar absorption for photocatalysis with black hydrogenated titanium dioxide nanocrystals[J]. Science, 2011, 331:746-750.

[96] TADA H, MITSUI T, KIYONAGA T, et al. All-solid-state z-scheme in CdS-Au-TiO$_2$ three-component nanojunction system[J]. Nat. Mater., 2006, 5:782-786.

[97] LEE H J, BANG J, PARK J, et al. Multilayered semiconductor (CdS/CdSe/ZnS)-sensitized TiO_2 mesoporous solar cells: all prepared by successive ionic layer adsorption and reaction processes[J]. Chem. Mater., 2010, 22:790-784.

[98] WANG W K, CHEN J, LI W W, et al. Synthesis of Pt-loaded self-interspersed anatase TiO_2 with a large fraction of {001} facets for efficient photocatalytic nitrobenzene degradation[J]. ACS Appl. Mater. Interfaces, 2015, 7:20349-20359.

[99] PANY S, NAIK B, MARTHA S, et al. Plasmon induced nano Au particle decorated over S, N-modified TiO_2 for exceptional photocatalytic hydrogen evolution under visible light[J]. ACS Appl. Mater. Interfaces, 2014, 6: 839-846.

[100] HIRAKAWA T, KAMAT P V. Charge separation and catalytic activity of Ag@TiO_2 core-shell composite clusters under UV-irradiation[J]. J. Am. Chem. Soc., 2005, 127:3928-3934.

[101] TSUKAMOTO D, SHIRO A, SHIRAISHI Y, et al. Photocatalytic H_2O_2 production from ethanol/O_2 system using TiO_2 loaded with Au-Ag bimetallic alloy nanoparticles[J]. ACS Catal., 2012, 2: 599-603.

[102] BAKER D R, KAMAT P V. Photosensitization of TiO_2 nanostructures with CdS quantum dots: particulate versus tubular support architectures[J]. Adv. Funct. Mater., 2010, 19:805-811.

[103] SUN W T, YU A, PAN H Y, et al. CdS quantum dots sensitized TiO_2 nanotube-array photoelectrodes[J]. J. Am. Chem. Soc., 2009, 130:1124-1125.

[104] JIN S, LIAN T. Electron transfer dynamics from single CdSe/ZnS quantum dots to TiO_2 nanoparticles[J]. Nano Lett., 2009, 9:2448.

[105] WANG M, ANGHEL A M, MARSAN B T, et al. CoS supersedes Pt as efficient electrocatalyst for triiodide reduction in dye-sensitized solar cells [J]. J. Am. Chem. Soc., 2009, 131:15976-15977.

[106] ULLAH K, YE S, SARKAR S, et al. Photocatalytic degradation of methylene blue by NiS_2-graphene supported TiO_2 catalyst composites[J]. Asian J. Chem., 2013, 26:145-150.

[107] LIU J, JIN W, KU Z, et al. Aqueous rechargeable alkaline $Co_xNi_{2-x}S_2$/TiO_2 battery[J]. Acs Nano, 2015, 10:1007-1016.

[108] WANG K, YANG B, YU L, et al. Preparation of Ni_2P/TiO_2-Al_2O_3 and the catalytic performance for hydrodesulfurization of 3-methylthiophene [J]. Energy Fuels, 2009, 23:4209-4214.

[109] IWASHINA K, IWASE A, YUN H N, et al. Z-schematic water splitting into H_2 and O_2 using metal sulfide as a hydrogen-evolving photocatalyst and reduced graphene oxide as a solid-state electron mediator[J]. J. Am. Chem. Soc., 2015, 137:604-607.

[110] TAN L, CHAI S, MOHAMED A R. Synthesis and applications of graphene-based TiO_2 photocatalysts [J]. ChemSusChem, 2012, 5:1868-1882.

[111] WOAN K, PYRGIOTAKIS G, SIGMUND W. Photocatalytic carbon-nanotube-TiO_2 composites[J]. Adv. Mater., 2009, 21:2233-2239.

[112] OH W C, ZHANG F J, CHEN M L. Preparation of carbon nanotubes/TiO_2 composites with multi-walled carbon nanotubes and titanium alkoxides by solvent effect and their photocatalytic activity[J]. Asian J. Chem., 2010, 22:2231-2243.

[113] EDER D, WINDLE A H. Carbon-inorganic hybrid materials: the carbon-nanotube/TiO_2 interface[J]. Adv. Mater., 2008, 20:1787-1793.

[114] LI Q, GUO B, YU J, et al. Highly efficient visible-light-driven photocatalytic hydrogen production of CdS-cluster-decorated graphene nanosheets [J]. J. Am. Chem. Soc., 2011, 133:10878-10884.

[115] JIA L, WANG D H, HUANG Y X, et al. Highly durable N-doped graphene/CdS nanocomposites with enhanced photocatalytic hydrogen evolution from water under visible light irradiation[J]. J. Phys. Chem. C, 2011, 115:11466-11473.

[116] CHEN C, SHI H, ZHAO G. Chiral recognition and enantioselective photoelectrochemical oxidation toward amino acids on single-crystalline ZnO[J]. J. Phys. Chem. C, 2014, 118:12041-12049.

[117] WULFF G. Molecular recognition in polymers prepared by imprinting with templates[J]. ACS SYM SER, 1986, 5:186-230.

[118] VLATAKIS G, ANDERSSON L I, MüLLER R, et al. Drug assay using antibody mimics made by molecular imprinting[J]. Nature, 1993, 361:645-647.

[119] WULFF G. Molecular imprinting in cross-linked materials with the aid of

molecular templates: a Way towards artificial antibodies[J]. Angew. Chem. Int. Ed. 1995, 34: 1812-1832.

[120] YANG, ZHENGPENG, ZHANG, et al. Facile synthesis of magnetically recoverable Fe_3O_4/Al_2O_3/molecularly imprinted TiO_2 nanocomposites and its molecular recognitive photocatalytic degradation of target contaminant [J]. J. Mol. Catal. A-Chem., 2015, 402: 10-16.

[121] SHEN X, ZHU L, JING L, et al. Synthesis of molecular imprinted polymer coated photocatalysts with high selectivity[J]. Chem. Comm., 2007, 11: 1163-1165.

[122] XU S, LU H, CHEN L, et al. Molecularly imprinted TiO_2 hybridized magnetic Fe_3O_4 nanoparticles for selective photocatalytic degradation and removal of estrone[J]. RSC Adv., 2014, 4: 45266-45274.

[123] HONGTAO, WANG, XUAN, et al. Enhanced photocatalytic degradation of tetracycline hydrochloride by molecular imprinted film modified TiO_2 nanotubes[J]. Chinese Sci. Bull., 2012, 57: 601-605.

[124] SHEN X, ZHU L, WANG N, et al. Molecular imprinting for removing highly toxic organic pollutants[J]. Chem. Comm., 2012, 48: 788-798.

[125] LIU Y, LIU R, LIU C, et al. Enhanced photocatalysis on TiO_2 nanotube arrays modified with molecularly imprinted TiO_2 thin film[J]. J. Hazard. Mater., 2010, 182: 912-918.

[126] CHAI S, ZHAO G, ZHANG Y N, et al. Selective photoelectrocatalytic degradation of recalcitrant contaminant driven by an n-P heterojunction nanoelectrode with molecular recognition ability[J]. Environ. Sci. Technol., 2012, 46: 10182-10190.

[127] ZHANG Y N, DAI W G, WEN Y Z, et al. Efficient enantioselective degradation of the inactive (S)-herbicide dichlorprop on chiral molecular-imprinted TiO_2[J]. Appl. Catal. B., 2017, 212: 185-192.

[128] TANG B, SHI H J, FAN Z Y, et al. Preferential electrocatalytic degradation of 2,4-dichlorophenoxyacetic acid on molecular imprinted mesoporous SnO_2 surface[J]. Chem. Eng. J., 2018, 334: 882-890.

[129] LUO Y, LU Z, JIANG Y, et al. Selective photodegradation of 1-methylimidazole-2-thiol by the magnetic and dual conductive imprinted photocatalysts based on TiO_2/Fe_3O_4/MWCNTs[J]. Chem. Eng. J., 2014, 240: 244-252.

[130] SAMBUR J B, CHEN T Y, CHOUDHARY E, et al. Sub-particle reaction and photocurrent mapping to optimize catalyst-modified photoanodes[J]. Nature, 2016, 530:77-80.

[131] ZHENG Z, TACHIKAWA T, MAJIMA T. Single-particle study of Pt-modified Au nanorods for plasmon-enhanced hydrogen generation in visible to near-infrared region[J]. J. Am. Chem. Soc., 2014, 136:6870-6873.

[132] ZHENG Z, TACHIKAWA T, MAJIMA T. Plasmon-enhanced formic acid dehydrogenation using anisotropic Pd-Au nanorods studied at the single-particle level[J]. J. Am. Chem. Soc., 2015, 137:948-957.

[133] HENDERSON M A, LYUBINETSKY I. Molecular-level insights into photocatalysis from scanning probe microscopy studies on TiO_2 {110}[J]. Chem Rev., 2013, 113:4428-4455.

[134] TAN S, JI Y, ZHAO Y, et al. Molecular oxygen adsorption behaviors on the rutile TiO_2 {110}-1×1 Surface: an in situ study with low-temperature scanning tunneling microscopy[J]. J. Am. Chem. Soc., 2011, 133:2002-2009.

[135] TACHIKAWA T, YAMASHITA S, MAJIMA T. Evidence for crystal-face-dependent TiO_2 photocatalysis from single-molecule imaging and kinetic analysis[J]. J. Am. Chem. Soc., 2011, 133:7197-7204.

[136] YE Z, LIU H, WANG F, et al. Single-particle tracking discloses binding-mediated rocking diffusion of rod-shaped biological particles on lipid membranes[J]. Chem. Sci., 2019, 10:1351-1359.

[137] ZHANG D, WEI L, ZHONG M, et al. The morphology and surface charge-dependent cellular uptake efficiency of upconversion nanostructures revealed by single-particle optical microscopy[J]. Chem. Sci., 2018, 9:5260-5269.

[138] YE Z, WENG R, MA Y, et al. Label-free, single-particle, colorimetric detection of permanganate by GNPs@Ag core-shell nanoparticles with dark-field optical microscopy[J]. Anal. Chem., 2018, 90:13044-13050.

[139] ZHAO Y, WANG Z, CUI X, et al. What are the adsorption sites for CO on the reduced TiO_2 {110}-1×1 surface? [J]. J. Am. Chem. Soc., 2009, 131:7958-7959.

[140] TAN S, HAO F, JI Y, et al. Observation of photocatalytic dissociation of water on terminal Ti sites of TiO_2 {110}-1×1 surface[J]. J. Am. Chem.

Soc., 2012, 134:9978-9985.

[141] YANG W, SUN H, TAN S, et al. Role of point defects on the reactivity of reconstructed anatase titanium dioxide {001} surface[J]. Nature Commun., 2013. DOI: 10.1038/ncomms3214.

[142] HAASE G. Surface photovoltage imaging for the study of local electronic structure at semiconductor surfaces[J]. Int. Rev. Phys. Chem., 2000, 19: 247-276.

[143] KRONIK L, SHAPIRA Y. Surface photovoltage phenomena: Theory, experiment, and applications[J]. Surf. Sci. Rep., 1999, 37:1-206.

[144] ZHAO J, OSTERLOH F E. Photochemical charge separation in nanocrystal Pphotocatalyst films: insights from surface photovoltage spectroscopy[J]. J. Phys. Chem. Lett., 2014, 5:782-786.

[145] ZHAO, J., WANG, et al. Photochemical charge transfer observed in nanoscale hydrogen evolving photocatalysts using surface photovoltage spectroscopy[J]. Energy Environ. Sci., 2015, 8:2790-2796.

[146] CHEN R, FAN F, DITTRICH T, et al. Imaging photogenerated charge carriers on surfaces and interfaces of photocatalysts with surface photovoltage microscopy[J]. Chem. Soc. Rev., 2018, 47:8238-8262.

[147] ZHU J, FAN F, CHEN R, et al. Direct imaging of highly anisotropic photogenerated charge separations on different facets of a single $BiVO_4$ Photocatalyst[J]. Angew. Chem. Int. Ed., 2015, 54:9111-9114.

[148] PARK J, YANG J, LEE G, et al. Single-molecule recognition of biomolecular interaction via kelvin probe force microscopy[J]. ACS Nano., 2011, 5:6981-6990.

[149] GAO Y, ZHU J, AN H, et al. Directly probing charge separation at interface of TiO_2 phase junction[J]. J. Phys. Chem. Lett., 2017, 8:1419-1423.

[150] CHEN R, PANG S, AN H, et al. Charge separation via asymmetric illumination in photocatalytic Cu_2O particles[J]. Nat. Energy, 2018, 3:655-663.

[151] ZHU Y, SALVADOR P A, ROHRER G S. Buried charge at the TiO_2/$SrTiO_3$ {111} interface and its effect on photochemical reactivity[J]. Acs Appl. Mater. Inter., 2017, 9:7843.

第 2 章

铂负载高暴露{001}晶面的锐钛矿二氧化钛合成及其光催化性能

2.1 高暴露{001}晶面的锐钛矿二氧化钛研究进展

金属氧化物半导体在能源储存和环境修复方面有广泛的应用前景,例如光催化产氢和水体中有机污染物的降解[1-2]。TiO_2[3]是应用较为广泛的光催化剂之一,具有非常优异的物理化学性能和光学特性[4-6],而由于TiO_2改性的光催化剂在能源采集和水体净化时可以直接利用太阳能而不会产生新的环境污染物,近年来更是引起极大关注[7-8]。由此,对TiO_2晶体表面基础氧化还原机制的分析有利于合理设计可应用于能源和环境方面的新型光催化剂[9]。在很长的一段时期内,许多科研工作者致力于寻求能提高TiO_2光催化效率的方法,例如金属和非金属掺杂和材料组成成分的调控。因此我们设想,将纳米结构的贵金属和半导体结合或者不失为一种可提升其催化性能的方法[1,10-11]。

在TiO_2表面沉积贵金属,光照下产生的光生电子会被贵金属俘获,这样可以有效抑制光生电子和空穴的复合率光催化效率,从而达到提高光催化效率的目的[12-13]。当光生电子与TiO_2接触时就会转移到金属粒子上,所以贵金属的费米能级(E_F)向导带偏移直到达到E_F和导带间的平衡[12]。除了电子俘获的机制外,贵金属粒子上由于等离子体激发的高能电子可以转移至TiO_2上,促进电子和空穴的分离,这样电子-空穴对就能用于光催化反应[14]。基于以上原理,可见光辐射导致的表面等离子体共振(SPR)效应[15]可以有效增强催化剂的性能。近来,Au纳米颗粒的表面等离子体共振已经用于可见光响应的光催化剂[16-21]。而且Au颗粒的表面等离子体共振光吸收被认为是决定产氢反应速率的重要因素之一[15]。对于Au-TiO_2,局部的表面等离子体共振使得电子-空穴对的分离得到加强,因此可见光辐射下的光催化效率得到提升,光催化降解亚甲基蓝得以实现[22]。当然,除了Au或者Ag可与TiO_2形成复合材料[23]外,Pt也可以作为电子阱沉积在TiO_2上,发挥增强电子-空穴对分离的作用,对有机污染物进行光催化降解[24-26]。除此之外,由于Pt的逸出功较大,可以在TiO_2-金属复合的区域形成肖特基势垒,促进界面多电子转移[27]。这样,光生电子从TiO_2的导带迁移至金属相中,阻碍了光生电子与空穴的复合,随之增强了光催化活性。另外,Pt的电子阱和Au的表面等离子体共振效应可以协同地增强光催化活性[28]。而负载 Pt 的 TiO$_2$ 催化剂在产氢和甲醇完全氧化上整体活性的提升，可以同时归结为 Pt 对光生电子的有效俘获和 Pt 本身固有的高催化活性[29]。

相对于金红石 TiO$_2$，锐钛矿 TiO$_2$ 纳米结构传输电子更容易，分离光生电子-空穴对的效率更高，所以在光电催化体系中一般具有更优异的性能[30-32]。但是，合成锐钛矿 TiO$_2$ 纳米结构往往步骤复杂。催化反应一般发生在 TiO$_2$ 表面，所以催化剂的催化活性相当依赖于晶体结构和暴露的晶面[33-35]。TiO$_2$ 锐钛矿 {001} 晶面的表面能为 0.90 J/m^2，在生长过程中，过高的表面能导致 {001} 晶面非常容易消失[36]，直到有研究者首次合成出 {001} 晶面暴露比例为 47% 的单晶锐钛矿 TiO$_2$[30]。锐钛矿 TiO$_2$ {101} 晶面的表面能为 0.43 J/m^2[37]，是最稳定且表面能最小的晶面。以综上所述为基础，我们可以通过合成暴露晶面不同的锐钛矿 TiO$_2$ 来系统地研究晶体表面不同对光催化活性的影响。例如，锐钛矿 TiO$_2$ {001} 面对亚甲基蓝的光降解效率要远高于 {101} 面[38]，但是在水溶液中，{101} 面选择性转化丙三醇为羟基乙醛的光催化活性要明显高于 {001} 面[39]。这些矛盾的结论提醒我们锐钛矿 TiO$_2$ 不同晶面的光催化活性不仅受表面能的影响，同时也受目标污染物的影响[40]。而要进行晶面依赖的活性研究，最常研究的形貌就是 {001} 面和 {101} 面同时存在的截断的四方双锥锐钛矿 TiO$_2$[37]。

尽管以往的研究已经展现过 TiO$_2$ 在光催化反应中的晶面依赖活性[41-44]，但是以贵金属复合的 TiO$_2$ 作为研究对象，特别是在污染物降解系统中却很少见。同时，纳米晶体 TiO$_2$ 光催化降解硝基苯也已经被许多小组研究过[45-47]。例如，Pt/TiO$_2$ 上负载石墨烯[45]或者包覆二羟基萘[47]可以有效提高纳米 TiO$_2$ 对硝基苯的降解效率。此外，对 TiO$_2$ 进行表面修饰也可以选择性地增强污染物的转化效果[48]并促进晶面与目标污染物的相互作用[49-50]。将 Pt 纳米粒子选择性地沉积在 {101} 面上，光催化产氢反应和光催化降解甲基橙的过程都得到加强[51]。而将 Pt 纳米粒子同时沉积在 {101} 面和 {010} 面上，苯酚的降解效率可以达到最高[52]。然而，对于甲醛氧化为 CO$_2$ 的过程，Pt 复合的 TiO$_2$ 优势晶面的光催化活性却并未给出[53]。因此，我们还需要更多的研究去探索相应晶面对目标污染物的光氧化和光还原活性。在原子水平上理解复合的贵金属和晶体表面性能之间的关系，可以为我们提高锐钛矿 TiO$_2$ 在实际应用中的光催化活性提供理论依据。除此之外，通过探究晶面依赖的光催化反应，形貌规则的超晶格结构性能可得到提升[54]。对不同污染物在特定晶面上降解机制的系统性研究能够提供增强光催化活性的方法，从而降低成本，有利于大规模的应用。

我们采用第一性原理[55]，计算 Pt 负载 TiO$_2$ 的晶面对光催化性能产生的影响，发现 TiO$_2$ 的 {001} 面倾向于积累更多带正电的空穴，因而比 {101} 晶面具有更高的光催化活性。受此结果启发，我们致力于研究污染物在 Pt 负载 TiO_2 复合物的不同晶面上的降解机制，选择一个广泛应用的模型污染物硝基苯作为目标污染物来评估催化剂的光催化性能[56-57]。{001}面和{101}面共同暴露的多面体锐钛矿 TiO_2 纳米晶体是以 Ti 为源，并加入形貌调控剂 NH_4F 合成的，而且我们也探究了多面体的合成机制[58]。为了确定 Pt/TiO_2 中 TiO_2 暴露晶面不同对硝基苯降解效率的影响，我们合成了暴露晶面不同的 TiO_2 作为 Pt 纳米颗粒的基底。同样，我们也合成了不同形貌的 Pt 纳米颗粒以探究形貌不同带来的影响。

2.2 铂负载高暴露{001}晶面的锐钛矿二氧化钛的合成与表征

2.2.1

铂负载高暴露{001}晶面的锐钛矿二氧化钛的制备

将 40 mL 浓度为 6 mol/L 的 HCl 溶液与 0.4 g 钛片混合，水热法制成 TiO_2 纳米晶体，随后取出 15 mL 该溶液，加入 0.15 g NH_4F 并转移至 50 mL 烧杯。随后，将混合溶液搅拌 15 min，加入 5 mL 水，继续搅拌 15 min 后转移至不锈钢高压釜中的 50 mL 聚四氟乙烯内胆中。水热过程再电热烘箱 160 ℃条件下保持 12 h。反应完成后，使高压釜冷却至室温，取出釜中的样品，在超声条件下依次用无水乙醇和去离子水洗涤样品。在进行光催化性能测试前，所有的样品都用 0.1 mol/L 的 NaOH 溶液和去离子水冲洗干净以去除吸附在样品上的氟离子，所有反应中用到的试剂都为分析纯且未经过进一步纯化。

质量分数为 1% 的 Pt 纳米颗粒随后通过化学沉积的方法沉积在 TiO_2 的{001}面和{101}面，而将 Pt 纳米颗粒仅沉积在 TiO_2 的{101}面则采用光致还原反应。具体的沉积过程如下：首先，将 0.2 g TiO_2 固体，100 mL 乙醇，$H_2PtCl_6 \cdot 6H_2O$(2 mg Pt)和 50 mg 抗坏血酸加入三颈烧瓶中，在不停搅拌的条件下进行加热回流。经 1 h 的沉积后，得到灰色的固体产物，随后进行离心，60 ℃ 条件下干燥 1 h，最后在高温管式炉空气气氛中 450 ℃ 退火 1 h。

2.2.2

高暴露{001}晶面的锐钛矿二氧化钛的形成机制

首先，我们使用热场发射扫描电子显微镜(Zeiss 公司，德国)，在 5 kV 的加速电压下获得样品的扫描电子显微图像。锐钛矿 TiO_2 为切去顶端的八面体双锥，侧面为{101}面，顶部和底面均为{001}面[30,37]。图 2.1 为我们提出的自分散纳米结构的形成机制以及对应的中间态和最终锐钛矿 TiO_2 纳米晶体的 SEM 图。图 2.1(i~l)为放大率较小的大范围 SEM 图。如图 2.1(e)所示，水热开始反应 1.5 h 后，自分散的二维 TiO_2 纳米片就已经形成；随后，二维 TiO_2 纳米片开始相互叠加形成更加复杂的自分散结构(图 2.1(b、f、j))。在水热反应 3 h 后，得到表面附有 TiO_2 纳米小颗粒的三维块状 TiO_2 纳米结构(图 2.1(c、g、k))。最后，经过 12 h 水热反应之后，三维自组装的锐钛矿 TiO_2 结构形成，且表面有大部分光滑的{001}面。为此，我们以表面自由能和表面张力为依据，建立出热力学模型来评估表面化学在锐钛矿 TiO_2 纳米结构形成过程中对其形貌和相稳定性的影响[59-60]。根据这个模型，在锐钛矿 TiO_2 的合成过程中，我们能够对其进行裁剪形成所需要的形貌和暴露晶面。除此之外，多面体三维自分散的结构表现出独一无二的协同性能，自组装形成 TiO_2 纳米晶之间形成空隙，可以加速反应物和产物的扩散，提高光催化效率[61-62]。

根据杨化桂等人所发表的研究结果[30]，在合成暴露{001}晶面占主导地位的锐钛矿 TiO_2 晶体过程中，氟离子(F^-)起到了非常重要的作用。F—Ti 键较高的键能在很大程度上降低了{001}晶面的表面能[63]。在传统方法中，通常加入氢氟酸作为氟离子的来源，但是，我们加入了 NH_4F 来控制暴露的晶面。这样可以避免使用高腐蚀性的液体，同时，NH_4^+ 和 F^- 之间存在位阻效应和静电吸引力，有助于形成分散的纳米结构。此外，NH_4^+ 还可以在 TiO_2 表面自动键和，形成表面电偶极矩[64]。NH_4^+ 在合成过程中的重要作用给我们提供了进行进一步研究的契机。

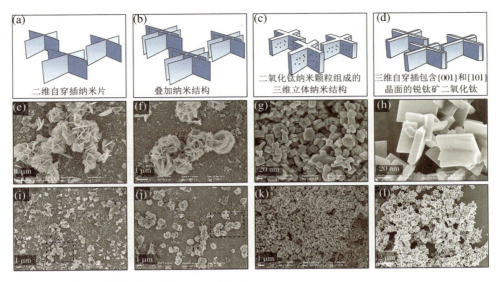

图 2.1 三维自穿插的 TiO₂ 可能的合成机制

(e、i) SEM 图显示为在水热 160 ℃,1.5 h 合成的二维自穿插纳米片;(f、j) 在水热 2 h 后自组装形成纳米结构;(g、k) 水热 3 h 后形成三维实心纳米结构且表面有部分小颗粒;(h、l)在水热 12 h 后形成高暴露{001}和{101}晶面的三维自穿插的锐钛矿 TiO₂

2.2.3

高暴露{001}晶面的锐钛矿二氧化钛的物相、微观结构及化学性质分析

在确认材料的形成过程后,我们使用 X 射线衍射仪(MXPAHF 公司,日本)对获得的样品进行 X 射线衍射分析,探究其物相构成(采用 Cu Kα 射线源,扫描速度为 8°/min,加速电压为 30 kV,管流为 3001mA)。图 2.2(a)显示了所合成的 TiO$_2$ 的 XRD 结果,证实我们合成的确实是锐钛矿 TiO$_2$,与标准卡片 JCPDS 21-1272 对比,衍射峰为 25.32°,37.93°,48.02°,53.98°和 55.04°,分别对应于锐钛矿 TiO$_2$ 的{101}、{004}、{200}、{105}和{211}晶面。此外,Pt/TiO$_2$ 复合物的衍射峰值显示光沉积在 TiO$_2$ 上的 Pt 为立方相(JCPDS 4-802)。

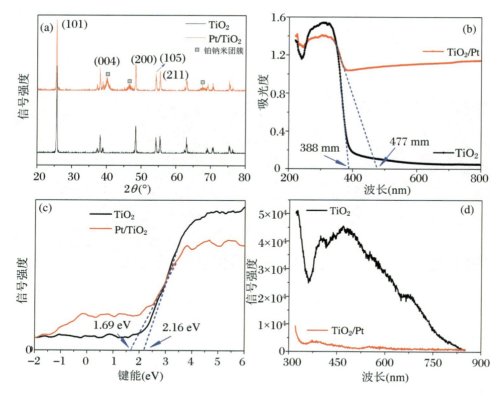

图 2.2 (a) XRD 图;(b) UV-vis 吸收谱;(c) XPS 价带谱;(d)纯 TiO_2 和 Pt 负载 TiO_2 的荧光光谱

图 2.3 中,SEM 和 TEM 图表明 TiO_2 颗粒为{001}面和{101}面围成的十面体,图 2.3(c)中经过还原沉积在 TiO_2 上的 Pt 纳米粒子均匀地分布在晶面上。为进一步直接观察 TiO_2 纳米晶体上的 Pt 团簇位置,我们进行了 HRTEM 测试,使用 JEM-2010 型透射电子显微镜(JEOL 公司,日本),在 200 kV 的加速电压下获得 Pt/TiO_2 的透射电子显微图、高分辨透射电子显微图和选区电子衍射花样。在图 2.3(d)中,HRTEM 图显示单个 TiO_2 纳米晶体的晶面间距为 0.19 nm,正好与十面体 TiO_2 的{020}和{200}晶面间距符合。图 2.3(d)中嵌入的选取电子衍射(SEAD)图进一步确定我们得到了锐钛矿纳米晶体{020}面,原点和{200}面间的夹角为 90°。HRTEM 图中观察得到的晶面间距与 XRD 的结果基本吻合。另外,沉积在 TiO_2 表面的 Pt 纳米颗粒尺寸非常均一,在 TiO_2 的{001}和{101}面上都均匀分布。为了确认 Pt 在两种材料的负载量关系,Pt/TiO_2 中 Pt 元素的含量通过电感耦合等离子发射光谱(ICP-AES)测试得到(PerkinElmer 公司,美国)。ICP-AES 结果表明,在这两个面上负载的 Pt 总量为 11.77 μg/mL。

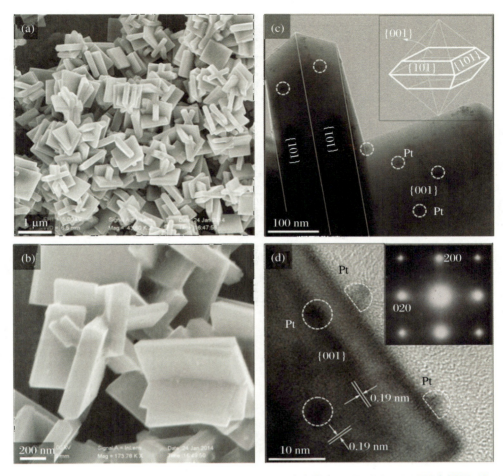

图2.3 (a) 水热合成 TiO_2 纳米晶形貌 SEM 图;(b) 放大的 SEM 图;(c) TEM 图(内插图:晶体轴向示意图);(d) Pt/TiO_2 样品 HRTEM 图(内插图(d):SAED 图)

为了排除样品的比表面积对材料光催化性能的影响,获得样品通过 BET 方法,在 TriStar Ⅱ 3020 V1.03 仪器上测得的比表面积(Micromeritics Instrument 公司,美国)。图 2.4 为采用 BET 测试得到的 Pt 负载锐钛矿 TiO_2 的相对比表面积和孔径分布。根据 BDDT 分类[65-66],吸附等温线属于Ⅳ型,磁滞回线为 H3 型,表明该材料为介孔材料。SEM 和 TEM 图显示中孔是由团聚在一起的纳米晶体形成,单独的纳米晶体内部却缺乏孔结构。经 BET 测试得到的 N_2 吸附曲线计算出锐钛矿 TiO_2 和 Pt 负载锐钛矿 TiO_2 的相对比表面积分别为 5.9 m^2/g 和 6.6 m^2/g,具有更大比表面积的 Pt 负载锐钛矿 TiO_2 多面体结构更有利于促进催化剂对污染物的吸附,从而提升催化剂的光降解性能。

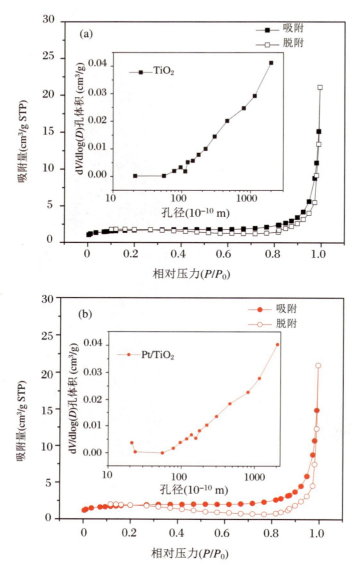

图 2.4 (a) 锐钛矿 TiO_2；(b) Pt/TiO_2 样品的 N_2 吸附脱附曲线
内插图为相对应的孔径分布图

为了进一步分析材料的化学成分，X 射线光电子能谱仪被用来测定并分析材料表面的元素价态及价带位置（Thermo Fisher 公司，美国）。图 2.5 的 XPS 测试谱图所展示的为 Pt 负载锐钛矿 TiO_2 纳米结构的表面组成和键能信息。由图 2.5(a) 谱图中包含 O、Ti 和 Pt 元素的峰，说明混合物表面组成为 Pt/TiO_2。谱图中 C 的 1s 峰来自于背底，图 2.5(b) 中显示 Ti 的 2p 谱由两个峰位分别为 458.5 eV 和 464.2 eV 的峰组成，分别对应于 Ti 的 $2p_{3/2}$ 峰和 $2p_{1/2}$ 峰，O 的 1s 谱峰位为 529.7 eV，来源于晶体表面有 Pt 沉积的 TiO_2 的 Ti—O 键，属于 O^{2-}。在

70.3 eV 和 73.6 eV 处的两个峰则分别对应于 Pt/TiO$_2$ 中 Pt 的 4f$_{7/2}$ 峰和 4f$_{5/2}$ 峰,图 2.5 显示其分裂能 Δ = 3.3 eV,表明催化剂中含有金属 Pt[67]。

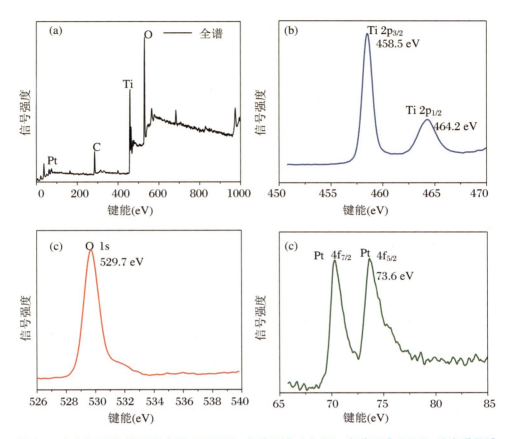

图 2.5 (a) Pt/TiO$_2$ 的 XPS 全谱;(b) Ti 2p 光谱区域;(c) O 1s 光谱区域;(d) Pt 4f 光谱区域

相应样品的漫反射光谱(DRS)通过 UV-vis 分光光度计测量(Shimadzu 公司,日本)。图 2.2(b)为催化剂的紫外-可见漫反射谱(DRS),可见在引入 Pt 后,在可见光波段吸收明显增强,并且 Pt/TiO$_2$ 复合物比单纯的 TiO$_2$ 纳米晶体表现出更高的吸收率,这可以归结为 Pt 纳米粒子的电子阱效应[15],催化剂的光学禁带可以通过以下公式[68]计算得到:

$$\alpha(h\nu) = A(h\nu - E_g)^{n/2} \quad (2.1)$$

其中,α 为吸收系数,$h\nu$ 为光子能量,E_g 为光学禁带,直接跃迁情况下 $n = 1$(非直接跃迁情况 $n = 4$),A 是与跃迁几率有关的常数。

TiO$_2$ 和 Pt/TiO$_2$ 的吸收边的位置分别在 388 nm 和 477 nm 处(图 2.2(b)),波长数据可以转化为带隙值,分别为 3.2 eV 和 2.6 eV[69-70],每个样品的价带(VB)可以在图 2.2(c)的 XPS 价带谱中看出,Pt/TiO$_2$ 低于费米能级的价带边为 1.69 eV,表明 Pt 的存在造成了价带向低能量方向有轻微位移。另外,Pt/TiO$_2$

的导带为 -0.91 eV,位置比单独的 TiO_2 多面体纳米结构更靠近费米能级。这些结果都说明,Pt/TiO_2 在可见光区域有更窄的带隙,这与以往的研究[71-72]及我们的理论预测是一致的[55]。因此,这样负载 Pt 且暴露晶面为{001}面和{101}面的多面体 TiO_2 结构能够通过促进光生电子和空穴的分离从而更进一步地提高其催化活性。

另外,各个样品的光致发光(PL)通过荧光分光光度计测定(Shimadzu 公司,日本),获得的荧光谱图显示,在 Pt 存在的情况下的确降低了电子和空穴的复合率。正如图 2.2(d)中所展示的,在 300~850 nm 波段,Pt/TiO_2 的光致发光强度远低于纯 TiO_2。因为光致发光是光生电子和空穴复合所发射的荧光,所以 Pt/TiO_2 的光致发光强度降低正好说明电子和空穴复合中心密度的下降,相应地,反应中就有更多的光生电子产生作用。而电子和空穴复合减少可以归因于光生电子从 TiO_2 导带到 Pt 团簇的传导速度加快。

2.2.4
铂负载高暴露{001}晶面的光催化及转化效率

为了明确 Pt 纳米颗粒的具体作用,分别以 Pt/TiO_2 和单纯的 TiO_2 作为催化剂在相同条件下对硝基苯进行光降解实验,实验各重复了 3 次,入射可见光强度($I_{0,vis}$)大约为 160 mW/cm^2,紫外光初始强度($I_{0,UV}$)为 157 mJ/cm^2 每 60 s,也就是 2.6 mW/cm^2。以 Xe 灯(CHF-XM-350W,Beijing Trusttech 公司,中国)作为光催化降解实验的光源,紫外($\lambda<420$ nm)或可见光($\lambda\geqslant 420$ nm)光源则通过加入滤光片来获取。在实验开始前,向 25 mL 的反应溶液中加入合成的催化剂,使其浓度为 1 mg/mL,反应体系中硝基苯的初始浓度为 10 mg/L。在外部施加 15 A 的电流产生光辐射,诱导反应发生。在光降解的过程中,每隔 10 min 取样分析硝基苯的浓度和降解产物。利用 TOC 分析仪(Multi N/C 2100,Analytik 公司,德国)测试样品中的总碳浓度来确认硝基苯是否被矿化。可见光的光密度和紫外光的光密度分别利用辐射计(Model FZ-A,Photoelectric Instrument 公司,中国)和焦耳计(KUÜHNAST 公司,德国)进行测量。光转化效率可以根据外部施加一定电压下的光电流密度进行确定。采用配置有 Hypersil-ODS 反相柱和 VDW 检测器的高性能液相色谱(HPLC-1100,Agilent 公司,美国)测量硝基苯的浓度。流动相的成分为 0.1%乙酸和甲醇与水的体积比为 40∶60 的混合液,流动速率为 0.8 mL/min。

图 2.6(a)和图 2.6(b)展示了硝基苯随时间变化的光降解曲线,降解遵循一级动力学。因此,具体的硝基苯降解速率常数(k)可以通过 $\ln(C/C_0)$-t 拟合曲线的斜率求出,其中 C 代表硝基苯的浓度。误差值给出曲线的标准差,关于硝基苯光降解速率的详细结果列在表 2.1 中。将不同催化剂条件下得到的硝基苯降解速率常数的结果进行对比,发现在紫外光照情况下,以 Pt/TiO_2 为催化剂硝基苯降解速率最快,相同紫外照射情况下,k_{Pt/TiO_2} 绝对值是 k_{TiO_2} 绝对值的两倍。除此之外,紫外照射下,1 h 内以 Pt/TiO_2 为催化剂的体系中硝基苯的整体去除率比单纯 TiO_2 为催化剂的体系中的整体去除率高出 30%。所有得到的结果表明,Pt 的存在很大程度上提高了硝基苯的光降解效率。值得注意的是,在没有任何光催化剂的情况下硝基苯的浓度也有略微减少,可能是由于系统中存在直接的光降解。这种情况在其他污染物的降解体系中也有出现[45]。一般情况下,我们以 TOC 去除率来衡量硝基苯在光催化降解中的矿化率,在反应 1 h 后,以 Pt/TiO_2 为催化剂的体系中 TOC 去除率为 55%,而 TiO_2 为催化剂的体

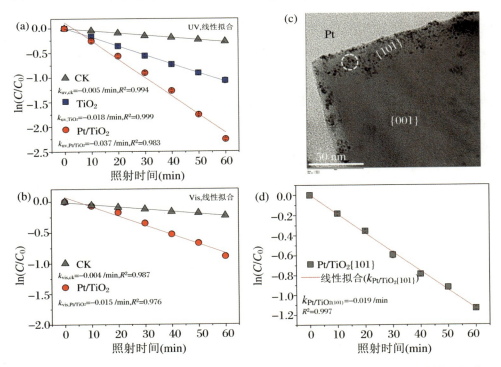

图 2.6 在紫外下光催化降解硝基苯实验($I_{0,UV}$ = 2.6 mW/cm^2):(a) 空白、单独的 TiO_2 和 Pt/TiO_2 作为催化剂;(b) 相同降解实验在可见光下($I_{0,vis}$ = 160 mW/cm^2);(c) Pt/TiO_2 {101}样品的 TEM 图;(d) 在紫外下光催化降解硝基苯实验,Pt/TiO_2 {101}作为催化剂($I_{0,UV}$ = 2.6 mW/cm^2)。

C_0 和 C 分别代表在 t = 0 和 t = t 时的硝基苯浓度

系中 TOC 去除率为 46%。结果说明，Pt/TiO$_2$ 有更高的矿化能力，主要是因为 Pt 纳米粒子促进了电子和空穴的分离。

表 2.1　可见光和紫外光下光催化降解硝基苯的动力学常数和回归系数

	光催化剂	斜率动力学常数 (/min)		统计
		数值	标准误差	R^2
可见光	Pt/TiO$_2$	-0.015	0.00095	0.976
紫外光	TiO$_2$	-0.018	0.00025	0.999
	Pt/TiO$_2$	-0.037	0.00203	0.983
紫外光	Pt/TiO$_2${101}	-0.019	0.00045	0.997

在光催化过程中，光转化率（ε_{eff}）反映了催化剂的实际太阳光的利用效率。我们采用电化学工作站 CHI 760D（Shanghai Chenhua 公司，中国）测量反应中的光电流密度。在该三电极体系中，电解质溶液含有浓度为 1 mg/mL 的光催化剂，修饰过的 FTO 导电玻璃为工作电极，Pt 丝为对电极，Ag/AgCl 电极为参比电极。所有的光电流测试均是在 0.1 mol/L 的 Na$_2$SO$_4$ 溶液（pH=7.0）中进行，外加电压（E_{app}）为 0.1 V（0.1 V 是以 Ag/AgCl 为标准，换算成标准氢电势为 0.3 V）。

$$E_{app} = E_{meas} - E_{aoc} \qquad (2.2)$$

式中，E_{meas} 为测试光电流时对应的电极电势，E_{aoc} 为同样测试条件下的开路电势[58]。

光转化率（ε_{eff}）反映了在不同波段的量子效率（QE）平均值[58,73]，QE 对于半导体，表示的是一个材料内从激子产生到发射光子的绝对产出，和材料固有的光子和光电特性有关，有外加电压的情况下，光转化率可以表示为[52,58]

$$\varepsilon_{eff} = j_p(E_{rev}^{\ominus} - |E_{app}|)/I_0 \qquad (2.3)$$

式中，j_p 是光电流密度（$\mu A/cm^2$），$E_{rev}^{\ominus}$ 为反应发生时的热力学标准态可逆电势，I_0 为光强密度（mW/cm^2），$|E_{app}|$ 为施加电压绝对值，水裂解反应的热力学标准态可逆电势（$E_{rev}^{\ominus} = 1.23$ V）用来计算紫外或可见光条件下光催化剂的光转化率。

为进一步研究反应体系的光响应，图 2.7 中，紫外光密度和可见光密度为分别为 2.6 mW/cm^2 和 160 mW/cm^2，在光源开启和关闭循环间隔 100 s 和外加电压为 0.3 V（标准氢电势）的条件下，记录下随时间变化的光电流。由此得到在紫外光下的 $\varepsilon_{eff,Pt/TiO_2}/\varepsilon_{eff,TiO_2}$ 比值约为 1.3，而 $\varepsilon_{eff,Pt/TiO_2,UV}/\varepsilon_{eff,Pt/TiO_2,vis}$ 比值已接近 600。包含更多细节的光转化效率比值按如下方式计算得到。

在入射光强度相同时，$\varepsilon_{\text{eff,Pt/TiO}_2}/\varepsilon_{\text{eff,TiO}_2}$ 比值可以表示为

$$\frac{\varepsilon_{\text{eff,Pt/TiO}_2}}{\varepsilon_{\text{eff,TiO}_2}} = \frac{j_{\text{p,Pt/TiO}_2}}{j_{\text{p,TiO}_2}} \tag{2.4}$$

如图 2.7(a)，紫外光下的 $j_{\text{p,Pt/TiO}_2}$ 值为 $(0.0657 \pm 0.0063)\ \mu\text{A/cm}^2$，$j_{\text{p,TiO}_2}$ 值为 $(0.0521 \pm 0.0027)\ \mu\text{A/cm}^2$。因此，入射光强为 $2.6\ \text{mW/cm}^2$ 时，$\varepsilon_{\text{eff,Pt/TiO}_2}/\varepsilon_{\text{eff,TiO}_2}$ 比值为 1.3。

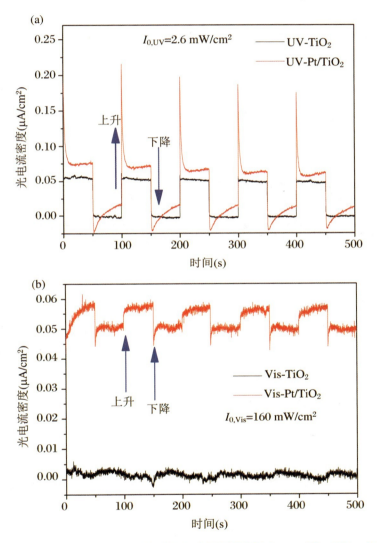

图 2.7 在紫外下 ($I_{0,\text{UV}} = 2.6\ \text{mW/cm}^2$) (a)；在可见光下 ($I_{0,\text{vis}} = 160\ \text{mW/cm}^2$) (b)；制备的 TiO_2/FTO 和 Pt/TiO_2/FTO 电极分别对应的光电流响应，在 $0.1\ \text{mol/L}\ Na_2SO_4$ 溶液中，加 $0.3\ \text{V}$ 偏压 (vs$_0$ SHE)，灯的开关循环为 $100\ \text{s}$

当入射光强不固定时，$\varepsilon_{\text{eff,Pt/TiO}_2,\text{UV}}/\varepsilon_{\text{eff,Pt/TiO}_2,\text{vis}}$ 比值可以表示为

$$\frac{\varepsilon_{\text{eff},\text{Pt}/\text{TiO}_2,\text{UV}}}{\varepsilon_{\text{eff},\text{Pt}/\text{TiO}_2,\text{vis}}} = \frac{j_{\text{p},\text{Pt}/\text{TiO}_2,\text{UV}} \cdot I_{0,\text{vis}}}{j_{\text{p},\text{Pt}/\text{TiO}_2,\text{vis}} \cdot I_{0,\text{UV}}} \tag{2.5}$$

光强为 160 mW/cm² 的可见光下的 $j_{\text{p},\text{Pt}/\text{TiO}_2}$ 值为 $(0.068 \pm 0.0002)\ \mu\text{A/cm}^2$，因此 $\varepsilon_{\text{eff},\text{Pt}/\text{TiO}_2,\text{UV}}/\varepsilon_{\text{eff},\text{Pt}/\text{TiO}_2,\text{vis}}$ 比值为 591。

这些结果证实了 Pt/TiO_2 确实使光致发光的强度减小，金属 Pt 在金属-半导体界面促进电荷分离从而增强了光催化效率[16]。Pt 颗粒由于比 TiO_2 有更强的电子亲和力，可以作为电子阱保留来自 TiO_2 的光生电子；因此阻碍电子与空穴的复合，最终达到提升光催化效率的目的。

2.3 铂负载高暴露{001}晶面的锐钛矿二氧化钛的光催化硝基苯机制分析

为了更好地理解在 Pt/TiO_2 上光催化转化硝基苯的机制，如图 2.8 所示，我们采用了液相色谱质谱联用仪（LC/MS，Agilent 公司，美国）分析反应中的中间产物。在光照 1 h 后，硝基苯的浓度下降了 90%，但是 LC/MS 只检测到了硝基酚这一种中间产物，说明降解过程是通过 ·OH 氧化进行的。水溶液中的羟基在价带带正电的空穴（h_{VB}^+）处反应产生 ·OH。根据以前的研究[74]，还原路径下硝基苯降解的中间产物包括亚硝基苯和苯基羟胺，最终得到的产物为苯胺，而硝基苯光催化氧化降解则主要是通过产生 ·OH[74]。此外，锐钛矿 TiO_2 的{001}面为光催化氧化位点，但是{101}晶面在光催化反应中却是还原位点[75]。氧化反应和还原反应在空间上被 TiO_2 的不同晶面分开，这表明，硝基苯的降解可能是在 TiO_2 的{001}晶面上经氧化路径发生，这与我们之前研究中的理论预测一致[55]。

为验证上述的结果，我们探索了硝基苯在 Pt/TiO_2{101}晶面上的光催化降解反应机制。仅在暴露的{101}晶面上负载 Pt 的锐钛矿 TiO_2 是通过光还原反应合成的，如图 2.6(c)所示，质量分数为 1% 的 Pt 纳米颗粒选择性地沉积在 TiO_2 晶体的{101}面上。之前在{001}面和{101}面上沉积的 Pt 总量为 11.77

μg/mL，但在 Pt/TiO₂{101} 面上沉积的 Pt 量就为 11.97 μg/mL，几乎与前一 Pt/TiO₂ 样品中的 Pt 总量相等。因此，可能由于负载 Pt 量不同对光催化性能造成的影响可以忽略。图 2.6(d) 展现的是在 Pt/TiO₂{101} 面上发生的硝基苯降解过程随时间变化的曲线。根据曲线可以看出，该反应也为一级动力学反应，表 2.1 为该光催化反应速率的具体数据。由此即可得到硝基苯降解的表观速率常数 k。结果显示，紫外光照下 $k_{Pt/TiO_2\{101\}}$ 的绝对值仅为 k_{Pt/TiO_2} 绝对值的一半。另外，1 h 紫外光照下的硝基苯总的去除率基本与以单纯 TiO₂ 为催化剂的体系总去除率相当，比以 Pt/TiO₂ 为催化剂的体系总去除率低。因此，仅仅将 Pt 团簇沉积在 TiO₂ 的 {101} 面对硝基苯的降解没有促进作用。这进一步说明，硝基苯的光催化降解更倾向于在负载 Pt 的锐钛矿 TiO₂ 的 {001} 表面通过氧化途径进行。

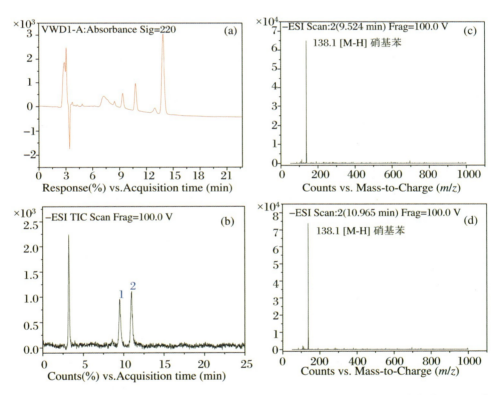

图 2.8　Pt/TiO₂ 光催化降解硝基苯过程中间产物的 LC/MS 谱图：(a) UV 吸收色谱图；(b) 液相谱图；(c) 9.524 min 时的中间产物质谱图；(d) 10.965 min 时的中间产物质谱图，LC/MS 条件；柱子：Zic-HILIC(150 mm × 2.0 mm i.d.)，流速：0.2 mL/min，流动相：甲醇/0.05% 甲酸 = 50/50(体积比)，柱温：30 ℃，进样量：1.0 μL，检测器：MS-ESI-m/z(10～300)

在本章中，我们提供了一种简捷绿色的高暴露 {001} 晶面锐钛矿 TiO₂ 的合成方法，并采用化学还原沉积法将 Pt 纳米颗粒均匀地负载在 TiO₂ 的 {001} 面和

{101}面上。所制得的 Pt/TiO$_2$ 催化剂用于硝基苯的光催化降解,反应 1 h 后 TOC 去除率为 55%,高于 TiO$_2$ 为催化剂的 46% TOC 去除率,说明 Pt/TiO$_2$ 具有更高的光催化活性和矿化能力,这主要是因为 Pt 纳米粒子作为助催化剂促进了电子和空穴的分离。·OH 产生于 Pt/TiO$_2$ 上的 {001} 晶面上,并用于光催化降解硝基苯。高暴露 {001} 晶面的锐钛矿 TiO$_2$ 通过 Pt 纳米颗粒的助催化作用,共同提高了对硝基苯光催化降解效率。该工作为设计降解目标污染物的光催化剂提供了一种新的思路,即通过控制暴露高能晶面和负载贵金属可以改善催化剂的活性,且该方法也适用于将其他贵金属纳米颗粒沉积在半导体催化剂上,以增强催化活性。

参考文献

[1] K S T, JANG Y H, K D H. A study on the mechanism for the interaction of light with noble metal-metal oxide semiconductor nanostructures for various photophysical applications[J]. Chem. Soc. Rev., 2013, 42: 8467-8493.

[2] LI X, ZHENG W, HE G. Morphology control of TiO$_2$ nanoparticle in microemulsion and its photocatalytic property[J]. ACS Sustainable Chem. Eng., 2013, 2: 288-295.

[3] GAO D D, LIU W J, XU Y. Core-shell Ag@Ni cocatalyst on the TiO$_2$ photocatalyst: one-step photoinduced deposition and its improved H$_2$-evolution activity[J]. Appl. Catal. B: Environ., 2020, 260: 8.

[4] LANG X J, MA W H, CHEN C C. Selective aerobic oxidation mediated by TiO$_2$ photocatalysis[J]. J. Am. Chem. Soc., 2014, 47: 355-363.

[5] HONG Y, JING X, HUANG J. Biosynthesized bimetallic Au-Pd nanoparticles supported on TiO$_2$ for solvent-free oxidation of benzyl alcohol[J]. ACS Sustain. Chem. Eng., 2014, 2: 1752-1759.

[6] NALDONI A, ALTOMARE M, ZOPPELLARO G. Photocatalysis with reduced TiO$_2$: from black TiO$_2$ to cocatalyst-free hydrogen production[J]. ACS Catal., 2019, 9: 345-364.

[7] CHEN C, MA W, ZHAO J J. Semiconductor-mediated photodegradation of pollutants under visible-light irradiation[J]. Chem. Soc. Rev., 2010, 39: 4206-4219.

[8] FUJISHIMA, NATURE H J. Electrochemical photolysis of water at a semi-

conductor electrode[J]. Nature., 1972, 238:37-8.

[9] KURNARAVEL V, MATHEW S, BARTLETT J. Photocatalytic hydrogen production using metal doped TiO_2: a review of recent advances[J]. Appl. Catal. B: Environ., 2019, 244:1021-1064.

[10] PU Y C, WANG G, CHANG K D. Au nanostructure-decorated TiO_2 nanowires exhibiting photoactivity across entire UV-visible region for photoelectrochemical water splitting[J]. Nano Lett., 2013, 13:3817-3823.

[11] JIANG X, FU X, ZHANG L. Photocatalytic reforming of glycerol for H_2 evolution on Pt/TiO_2: fundamental understanding the effect of co-catalyst Pt and the Pt deposition route[J]. J. Mater. Chem. A., 2015, 3:2271-2282.

[12] SUBRAMANIAN V, WOLF E E, KAMAT P V. Catalysis with TiO_2/gold nanocomposites. effect of metal particle size on the fermi level equilibration [J]. J. Am. Chem. Soc., 2004, 126:4943-4950.

[13] JAKOB M, LEVANON H, KAMAT P V J N L. Charge distribution between UV-irradiated TiO_2 and gold nanoparticles: determination of shift in the fermi level[J]. Nano Lett., 2003, 3:353-358.

[14] WU Y, LIU H, ZHANG J. Enhanced photocatalytic activity of nitrogen-doped titania by deposited with gold[J]. J. Phys. Chem. C., 2009, 113:14689-14695.

[15] TANAKA A, SAKAGUCHI S, HASHIMOTO K. Preparation of Au/TiO_2 with metal cocatalysts exhibiting strong surface plasmon resonance effective for photoinduced hydrogen formation under irradiation of visible light[J]. ACS Catal., 2013, 3:79-85.

[16] LINK S, CHRISTOPHER P, INGRAM D B J N M. Plasmonic-metal nanostructures for efficient conversion of solar to chemical energy[J]. Nat. Mater., 2011, 10:911-921.

[17] TIAN Y, TATSUMA T. Mechanisms and applications of plasmon-induced charge separation at TiO_2 films loaded with gold nanoparticles[J]. J. Am. Chem. Soc., 2005, 127:7632-7637.

[18] FURUBE A, DU L, HARA K. Ultrafast plasmon-induced electron transfer from gold nanodots into TiO_2 nanoparticles[J]. J. Am. Chem. Soc., 2007, 129:14852-14853.

[19] NAYA S I, TERANISHI M, ISOBE T. Light wavelength-switchable photocatalytic reaction by gold nanoparticle-loaded titanium(Ⅳ) dioxide[J].

Chem. Commun., 2010, 46:815-817.

[20] ZHOU X, LIU G, YU J. Surface plasmon resonance-mediated photocatalysis by noble metal-based composites under visible light[J]. J. Mater. Chem., 2012, 22:21337-21354.

[21] REDDY N L, RAO V N, VIJAYAKUMAR M. A review on frontiers in plasmonic nano-photocatalysts for hydrogen production[J]. I Int. J. Hydrog. Energy., 2019, 44:10453-10472.

[22] JUNQING, YAN, GUANGJUN. Synthetic design of gold nanoparticles on anatase TiO_2{001} for enhanced visible light harvesting[J]. ACS Sustainable Chem. Eng., 2014, 2:1940-1946.

[23] MENG A Y, ZHANG L Y, CHENG B. TiO_2-MnOx-Pt hybrid multiheterojunction film photocatalyst with enhanced photocatalytic CO_2-reduction activity[J]. ACS Appl. Mater. Interfaces., 2019, 11:5581-5589.

[24] KIM J, LEE J, CHOI W J C C. Synergic effect of simultaneous fluorination and platinization of TiO_2 surface on anoxic photocatalytic degradation of organic compounds[J]. Chem. Commun., 2008, 6(6):756-758.

[25] LEE J, CHOI W. Photocatalytic reactivity of surface platinized TiO_2, substrate specificity and the effect of Pt oxidation state[J]. J. Phys. Chem. B., 2005, 109: 7399-7406.

[26] WANG C Y, PAGEL R, BAHNEMANN D W. Quantum yield of formaldehyde formation in the presence of colloidal TiO_2-based photocatalysts: effect of intermittent illumination, platinization, and deoxygenation[J]. J. Phys. Chem. B., 2004, 108:14082-14092.

[27] TADA H, KIYONAGA T, NAYA S I J C S R. Rational design and applications of highly efficient reaction systems photocatalyzed by noble metal nanoparticle-loaded titanium(IV) dioxide[J]. Chem. Soc. Rev., 2009, 38:1849-1858.

[28] Facile photochemical synthesis of Au/Pt/g-C_3N_4 with plasmon-enhanced photocatalytic activity for antibiotic degradation[J]. ACS Appl. Mater. Interfaces., 2015, 7:9630-9637.

[29] NALDONI A, D'ARIENZO M, ALTOMARE M. Pt and Au/TiO_2 photocatalysts for methanol reforming: role of metal nanoparticles in tuning charge trapping properties and photoefficiency[J]. Appl. Catal. B: Environ., 2013, 130:239-248.

[30] YANG H G, SUN C H, QIAO S Z. Anatase TiO$_2$ single crystals with a large percentage of reactive facets[J]. Nature 2008, 453:638-641..

[31] YUAN S J, CHEN J J, LIN Z Q. Nitrate formation from atmospheric nitrogen and oxygen photocatalysed by nano-sized titanium dioxide [J]. Nat. Commun., 2013, 4:2249.

[32] ANANDAN S, RAO T N, SATHISH M. Superhydrophilic graphene-loaded TiO$_2$ thin film for self-cleaning applications[J]. ACS Appl. Mater. Interfaces., 2013, 5:207-212.

[33] HANSON K, LOSEGO M D, KALANYAN B. Stabilization of [Ru(Bpy)$_2$(4,4′-(PO$_3$H$_2$)Bpy)]$^{2+}$ on mesoporous TiO$_2$ with atomic layer deposition of Al$_2$O$_3$[J]. Chem. Mater., 2013, 25:3-5.

[34] LI X, YU J, LOW J. Engineering heterogeneous semiconductors for solar water splitting[J]. J. Mater. Chem. A., 2015, 3:2485-2534.

[35] CHEN Y J, JI S F, SUN W M. Engineering the atomic interface with single platinum atoms for enhanced photocatalytic hydrogen production [J]. Angew. Chem. Int. Ed., 2020, 59:1295-1301.

[36] LAZZERI M, VITTADINI A, SELLONI A. Structure and energetics of stoichiometric TiO$_2$ anatase surfaces[J]. Phys. Rev. B., 2001, 63:155409.

[37] LIU G, YANG H G, JIAN P. Titanium dioxide crystals with tailored facets [J]. Chem. Rev., 2014, 114:9559-612.

[38] HAN X, ZHENG B, OUYANG J. Control of anatase TiO$_2$ nanocrystals with a series of high-energy crystal facets via a fuorine-free strategy[J]. Chem. Asian J., 2012, 7:2538-2542.

[39] CHONG R, LI J, XIN Z. Selective photocatalytic conversion of glycerol to hydroxyacetaldehyde in aqueous solution on facet tuned TiO$_2$-based catalysts [J]. Chem. Commun., 2014, 50:165-167.

[40] CHEN W, KUANG Q, WANG Q. Engineering a high energy surface of anatase TiO$_2$ crystals towards enhanced performance for energy conversion and environmental applications[J]. RSC Adv., 2015, 5:20396-20409.

[41] ANGELIS F D, VITILLARO G, KAVAN L. Modeling ruthenium-dye-sensitized TiO$_2$ surfaces exposing the {001} or {101} faces: a first-principles investigation[J]. J. Phys. Chem. C., 2012, 116:18124-18131.

[42] LUAN Y, JING L, XIE Y. Exceptional photocatalytic activity of 001-facet-exposed TiO$_2$ mainly depending on enhanced adsorbed oxygen by residual

hydrogen fluoride[J]. ACS Catal., 2013, 3:1378-1385.

[43] ROY N, SOHN Y, PRADHAN D J A N. Synergy of low-energy {101} and high-energy {001} TiO$_2$ crystal facets for enhanced photocatalysis[J]. ACS Nano., 2013, 7:2532-2540.

[44] LI C, KOENIGSMANN C, DING W. Facet-dependent photoelectrochemical performance of TiO$_2$ nanostructures: an experimental and computational study[J]. J. Am. Chem. Soc., 2015, 137:1520-1529.

[45] JO W K, WON Y, HWANG I. Enhanced photocatalytic degradation of aqueous nitrobenzene using graphitic carbon-TiO$_2$ composites[J]. Ind. Eng. Chem. Res., 2014, 53:3455-3461.

[46] PATHAK T K, VASOYA N H, NATARAJAN T S. Photocatalytic degradation of aqueous nitrobenzene solution using nanocrystalline Mg-Mn ferrites [J]. Mater. Sci. Forum., 2013, 764:116-129.

[47] KAMEGAWA T, SETO H, MATSUURA S. Preparation of hydroxynaphthalene-modified TiO$_2$ via formation of surface complexes and their applications in the photocatalytic reduction of nitrobenzene under visible-light irradiation[J]. ACS Appl. Mater. Interfaces., 2012, 4:6635-6639.

[48] TADA H, ISHIDA T, TAKAO A. Drastic Enhancement of TiO$_2$-photocatalyzed reduction of nitrobenzene by loading Ag clusters[J]. Langmuir., 2004, 20:7898-7900.

[49] CROPEK D, KEMME P A, MAKAROVA O V. Selective photocatalytic decomposition of nitrobenzene using surface modified TiO$_2$ nanoparticles[J]. J. Phys. Chem. C., 2008, 112:8311-8318.

[50] MAKAROVA O V, RAJH T, THURNAUER M C. Surface modification of TiO$_2$ nanoparticles for photochemical reduction of nitrobenzene[J]. Environ. Sci. Technol., 2000, 34:4797-4803.

[51] LIU C, HAN X, XIE S. Enhancing the photocatalytic activity of anatase TiO$_2$ by improving the specific facet-induced spontaneous separation of photogenerated electrons and holes[J]. Chem. Asian J., 2013, 8:282-289.

[52] NI Y, WANG W. Facet-dependent photocatalytic mechanisms of anatase TiO$_2$: a new sight on the self-adjusted surface heterojunction[J]. J. Alloys Compd., 2015, 647:981-988.

[53] NIE L, YU J, LI X. Enhanced performance of NaOH-modified Pt/TiO$_2$ toward room temperature selective oxidation of formaldehyde[J]. Environ.

Sci. Technol., 2013, 47:2777-2783.

[54] ZHANG P, TACHIKAWA T, BIAN Z. Selective photoredox activity on specific facet-dominated TiO_2 mesocrystal superstructures incubated with directed nanocrystals[J]. Appl. Catal. B-Environ., 2015, 176:678-686.

[55] CHEN J J, WANG W K, LI W W. Roles of crystal surface in Pt-loaded titania for photocatalytic conversion of organic pollutants: a first-principle theoretical calculation [J]. ACS Appl. Mater. Interfaces., 2015, 7: 12671-12678.

[56] SUROLIA P K, TAYADE R J, JASRA R V J I. Photocatalytic degradation of nitrobenzene in an aqueous system by transition-metal-exchanged ETS-10 zeolites[J]. Ind. Eng. Chem. Res., 2010, 49:3961-3966.

[57] WAHAB H S, KOUTSELOS A. Computational modeling of the adsorption and (*)OH initiated photochemical and photocatalytic primary oxidation of nitrobenzene[J]. J. Mol. Model., 2009, 15:1237-1244.

[58] KHAN S U M, AL-SHAHRY M, INGLER W B. Efficient photochemical water splitting by a chemically modified n-TiO_2 [J]. Science., 2002, 297: 2243-2245.

[59] BARNARD A, CURTISS L A. Prediction of TiO_2 nanoparticle phase and shape transitions controlled by surface chemistry[J]. Nano Lett., 2005, 5: 1261-1266.

[60] BOURIKAS K, KORDULIS C, LYCOURGHIOTIS A J C. Titanium dioxide (anatase and rutile): surface chemistry, liquid-solid interface chemistry, and scientific synthesis of supported catalysts[J]. Chem Rev., 2015, 45: 9754-823.

[61] YU Y, CAO C, WEI L. Low-cost synthesis of robust anatase polyhedral structures with a preponderance of exposed {001} facets for enhanced photo-activities[J]. Nano Res., 2012, 5:434-442.

[62] ZHANG A Y, LONG L L, LI W W. Hexagonal microrods of anatase tetragonal TiO_2: self-directed growth and superior photocatalytic performance[J] Chem. Commun., 2013, 49:6075-6077.

[63] YU J, LOW J, XIAO W. Enhanced photocatalytic CO_2-reduction activity of anatase TiO_2 by coexposed {001} and {101} facets[J]. J. Am. Chem. Soc., 2014, 136:8839-8842.

[64] LIU G, TAO C, ZHANG M. Effects of surface self-assembled NH_4^+ on the

performance of TiO$_2$ based ultraviolet photodetectors[J]. J. Alloys Compd., 2014, 601:104-107.

[65] BRUNAUER S, DEMING L S, DEMING W E. On a theory of the van der waals adsorption of gases[J]. J. Am. Chem. Soc., 1940, 62:1723-1732.

[66] SING K J P, CHEMISTRY A. Reporting physisorption data for gas/solid systems with special reference to the determination of surface area and porosity (recommendations 1984)[J]. Pure Appl. Chem., 1985, 57: 603-619.

[67] LI Y H, XING J, CHEN Z J. Unidirectional suppression of hydrogen oxidation on oxidized platinum clusters[J]. Nat. Commun., 2013, 4:2500.

[68] YONGGANG, SHENG, AND. Low-temperature deposition of the high-performance anatase-titania optical films via a modified sol-gel route[J]. Opt. Mater., 2008, 30:1310-1315.

[69] WANG S, ZHAO L, BAI L. Enhancing photocatalytic activity of disorder-engineered C/TiO$_2$ and TiO$_2$ nanoparticles[J]. J. Mater. Chem. A., 2014, 2:7439-7445.

[70] FAN C, CHEN C, WANG J. Black hydroxylated titanium dioxide prepared via ultrasonication with enhanced photocatalytic activity[J]. Sci. Rep., 2015, 5:11712.

[71] SCHNEIDER J, MATSUOKA M, TAKEUCHI M. Understanding TiO$_2$ photocatalysis: mechanisms and materials[J]. Chem. Rev., 2014, 114: 9919-9986.

[72] ZHANG Z, YATES J T. Band bending in semiconductors: chemical and physical consequences at surfaces and interfaces[J]. Chem. Rev., 2012, 112:5520-5551.

[73] LUBER E J, BURIAK J M J A N. Reporting performance in organic photovoltaic devices[J]. ACS Nano., 2013, 7:4708-4714.

[74] ZHANG S J, JIANG H, LI M J. Kinetics and mechanisms of radiolytic degradation of nitrobenzene in aqueous solutions[J]. Environ. Sci. Technol., 2007, 41:1977-1982.

[75] TACHIKAWA T, YAMASHITA S, MAJIMA TJJOTACS. Evidence for crystal-face-dependent TiO$_2$ photocatalysis from single-molecule imaging and kinetic analysis[J]. J. Am. Chem. Soc., 2011, 133:7197-7204.

第 3 章

{001}晶面刻蚀蓝色锐钛矿二氧化钛微米球的合成及其光催化性能

3.1
高暴露{101}晶面的含氧空位锐钛矿二氧化钛研究进展

基于高暴露{001}晶面的 TiO_2 光催化性能研究[1-2]，晶面是控制 TiO_2 光催化性能的重要因素之一[3-5]。最近一些研究表明晶面在提高 TiO_2 光催化性能方面扮演重要的角色[6]。一个新的"surface heterojunction"概念在 DFT 计算的基础上被提了出来，它是为了强调光生电子和空穴在{101}和{001}晶面间的转移对 TiO_2 光催化性能的影响[7]。荧光单分子成像和动力学分析进一步确定了反应位点是在 TiO_2 的{101}晶面，而不是{001}晶面[8]。同时，在 TiO_2 晶面的电荷分离已经通过选择性的刻蚀{001}晶面得到证实[9-10]。在这种情况下，适当的刻蚀{001}晶面能够促进电子-空穴的分离和转移。此外，各种实验已经证明 TiO_2 的{101}晶面具有更高的热力学稳定性，而且在光产氢方面比{001}晶面表现出更高的活性[11-12]。

最近，已有文献报道包含 Ti^{3+} 的 TiO_2(TiO_{2-x})能够原位产生氧空位，这有利于电子的迁移[13]。与传统的掺杂相比，氧空位是一种没有引入杂原子的自掺杂方式。目前，已经有一些合成还原态 TiO_2 的方法报道。例如，利用 Zn 诱导合成 Ti^{3+} 掺杂的 TiO_{2-x}[14]。一些方法，例如等离子体处理[15]、激光辐射[16]或高能粒子轰击[17]也被用来在 TiO_{2-x} 中掺杂 Ti^{3+}。尽管 TiO_2 基材料在晶面和掺杂方面已经取得了很多进展，但仍希望找到能够一步实现 Ti^{3+} 的掺杂和优势晶面{101}暴露的方法。

在本章中，我们通过简单的水热法合成{101}晶面暴露的非化学计量的锐钛矿 TiO_2 微球；采用晶面刻蚀和掺杂的方法提高 TiO_2 微球的还原能力，通过紫外光照射下产氢来检测 TiO_2 的光催化能力是否得到提高；利用飞秒 TDR 进一步探索自掺杂的 Ti^{3+} 作为光生电子的捕获剂和{101}晶面的作用机制。

3.2 晶面刻蚀蓝色锐钛矿二氧化钛的合成与表征

3.2.1 晶面刻蚀蓝色锐钛矿二氧化钛的制备

本章工作中所用的化学试剂均为分析纯级别，且未作进一步纯化。Ti^{3+} 溶液是用之前报道的方法合成的[18-20]。简要的合成步骤如下：在 50 mL 烧杯中，将 0.15 g 的氟化铵溶于 15 mL Ti^{3+} 溶液中，搅拌 15 min 后，加入 25 mL 乙醇；再搅拌 15 min 后，将溶液转移至 50 mL 聚四氟乙烯水热釜中。在烘箱中 160 ℃ 下分别反应 4 h（标记为 TiO_2-1）和 12 h（标记为 TiO_2-2）。反应后自然冷却到室温。之后，将得到的样品一次用乙醇和水超声清洗干净。在光催化性能测试之前，所有的样品用 NaOH（0.1 mol/L）清洗来去除吸收的氟离子。TiO_2-3 是将样品 TiO_2-2 在 500 ℃ 下煅烧 4 h 得到的。

为制备 Pt/TiO_2 复合物，将 2 mg 的催化剂和一定量的 $H_2PtCl_6 \cdot 6H_2O$ 超声分散在 1 mL CH_3OH 和 4 mL H_2O 的混合溶液中，然后在磁力搅拌条件下光照（UV light 365 nm）0.5 h，得到的负载 Pt 的 TiO_2 样品用水清洗。

3.2.2 晶面刻蚀蓝色锐钛矿二氧化钛的物相及微观结构分析

首先，获得样品的 XRD 图谱通过 X 射线衍射仪（Rigaku 公司，日本）测试得到，使用 Cu Kα 放射线（λ=1.541841Å m）在 8°/min 条件下测试，加速电压和使用的电流分别为 40 kV 和 200 mA。图 3.1(a)中三个样品的 XRD 图谱具有相似的衍射峰，说明生成的样品是锐钛矿 TiO_2，没有其他物相的杂峰，与标准锐钛矿卡片（JCPDS No. 21-1272）一致。接着，TiO_2 纳米结构的形貌通过场发射扫描电子显微镜(FEI 公司,美国)和透射电子显微镜(JEOL 公司,日本)观测。图 3.1(b)和(c)中为不同水热时间获得的锐钛矿 TiO_2 微球的形貌。在图 3.1(b)和(c)中的插图能够明显观测到微观形貌分别是连续的纳米片和刻蚀的纳米片。刻蚀之后,在自组装的微球表面形成带和孔。如图 3.1(d)所示,样品 TiO_2-2 煅烧之后得到样品 TiO_2-3,形貌基本上没有发生变化,但是煅烧之后形成了一些小的颗粒。同时,Pt/TiO_2-2 样品的 XRD 图谱、SEM 图像和能量散射 X 射线光谱得到相应表征。

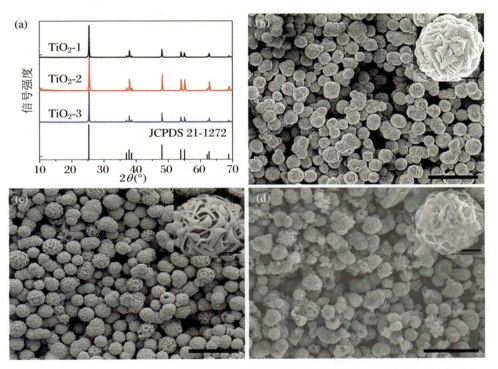

图 3.1 (a) 产物的 XRD 图;(b) TiO_2-1 反应 4 h 后的产物 SEM 图;(c) TiO_2-2 反应 12 h 后的产物 SEM 图;(d) TiO_2-3 样经过煅烧后的 SEM 图

标尺为 10 μm,内插标尺为 1 μm

3.2.3

晶面刻蚀蓝色锐钛矿二氧化钛的生长机制研究

为了进一步探究这个锐钛矿 TiO_2 三维分层结构的形貌发展和形成机制,我们进行了更多常规实验。图 3.2 是 TiO_2 微球的生长机制,是在 F^- 作用下合成的,水热温度为 433 K 时不同水热时间的样品。从系列样品的 SEM 图像得出,在起始反应阶段溶液中形成纳米片和纳米颗粒(图 3.2(a))。然后,纳米片和纳米颗粒聚集形成纳米球(图 3.2(b)),纳米球继续生长形成微球,暴露{101}和{001}晶面(图 3.2(c、d))[20]。

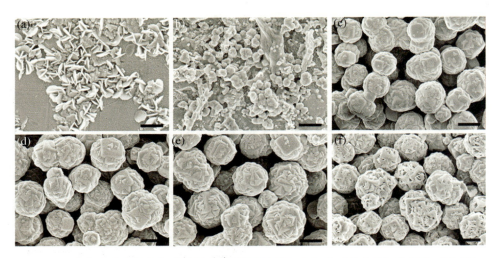

图 3.2　一步合成反应的各个时间的生长过程 SEM 图:(a) 1 h;(b) 1.5 h;(c) 2 h;(d) 3 h;(e) 4 h;(f) 6 h

标尺为 2 μm

在不同反应时间(3～6 h),产物 SEM 的图像(图 3.2(d～f))表明在图 3.1(c)中刻蚀得到的锐钛矿微球是由初始阶段的微球(图 3.2(d))选择性刻蚀得到的。由于反应过程中过量 HF 的刻蚀作用,大量的凹陷在锐钛矿 TiO_2 纳米片的连接处产生。这些凹陷处逐渐变大,而且{001}晶面的中心在刻蚀过程中几乎消失。结果产生了复杂的纳米结构(图 3.2(f))。有趣的是,当反应时间延长到 12 h,{001}晶面被完全刻蚀,得到更加复杂的结构。这样的一个形成过程可以看作原始结构的持续刻蚀[21]。在这个过程中,HF 在溶液中对锐钛矿 TiO_2 的{001}晶面具有刻蚀作用,而且刻蚀程度和产物的形貌与反应时间密切相关。

基于此,当 Pt 负载后,为了确认材料的形貌稳定性,使用 SEM 表征了 Pt 沉积的 TiO_2 样品,如图 3.3 所示。在 TiO_2-2 上沉积 Pt 后,TiO_2-2 样品的结构保持不变(图 3.3(a、b))。并且通过 EDS 测定,结果证实了在 TiO_2-2 的不同区域上存在负载的 Pt(图 3.3(c))。此外,制成样品的 XRD 图谱与标准锐钛矿的 XRD 图谱一致。由于负载的 Pt 含量低,未观察到 Pt 纳米颗粒的峰。

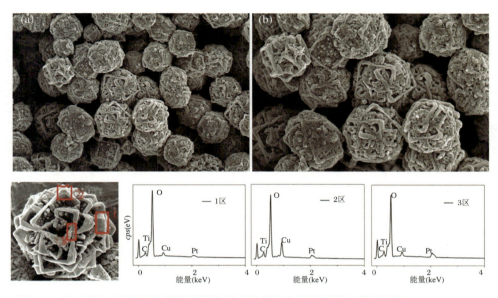

图 3.3　Pt／TiO$_2$-2 样品的 SEM 图像和相应的能量色散 X 射线光谱(EDS)

3.2.4
晶面刻蚀蓝色锐钛矿二氧化钛的光物理性质及光化学性质

为了探究材料的表界面缺陷,所获样品在 130 K 下测试得到的低温电子顺磁共振光谱测定(EPR,JEOL 公司,日本,140 K,9064 MHz,0.998 mW,X-band)。图 3.4 的低温 EPR 是在 130 K 下测试,用来说明在 TiO$_{2-x}$ 微球中存在 Ti^{3+}。在这个温度下,各向异性的 g 张量不能被解决[22-23]。TiO$_2$-2 在 $g=1.97$ 处有很强的 EPR 信号,这是由于次表面存在顺磁中心[24]。TiO$_2$-1 和 TiO$_2$-3 中 Ti^{3+} 的信号都非常弱,说明延长水热时间之后,随着腐蚀的加剧,在 TiO$_{2-x}$ 微球中生成大量 Ti^{3+} [25-26]。而在空气中煅烧过程中,TiO$_2$-3 样品中的 Ti^{3+} 能够被氧化为 Ti^{4+},氧空位减少[27-28]。

同时,EPR 图也排除了样品表面 Ti^{3+} 的存在[27]。因为如果表面存在 Ti^{3+} 则容易吸收空气中的 O$_2$,被还原为 O$_2^-$,在 $g=2.02$ 处出现 EPR 信号[28]。然而该样品在 $g=2.02$ 处没有观测到该信号[29]。

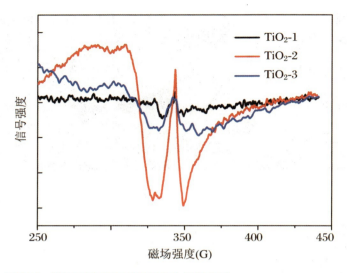

图 3.4 该获得样品在 130 K 下的 EPR 谱图

图 3.5(a、b)是刻蚀后微球的透射电子图像。图 3.5(c)是高分辨透射电子图像,插图是对应的选区电子衍射。图 3.5(a)的透射电子图像表明获得的 TiO_{2-x} 微球平均尺寸约为 3 μm,与图 3.2 中的扫描结果一致。在图 3.5(c)中清

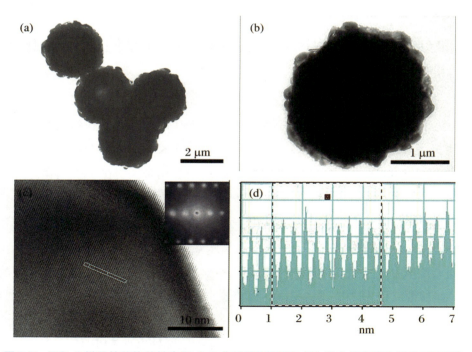

图 3.5 TiO_2-2 样品的结构特性表征:(a) 典型的 TEM 图;(b) 单个具有代表性的 TEM 图;(c) 表面刻蚀的微球样品的 HRTEM 图,内插图为选区电子衍射图;(d) 软件 Digital Micrograph 处理图

晰的晶格条纹说明刻蚀得到的 TiO_{2-x} 微球结晶性良好,晶面间距为 0.36 nm(图 3.5 为 Digital Micrograph 测量结果),与锐钛矿 TiO_2 的{101}晶面一致。更重要的是,为{001}晶面的刻蚀提供了直接证据,{101}刻蚀后保留下来。这与上述关于形貌的变化和形成机制的分析一致。

BET(Brunauer-Emmett-Teller)表面积的数据对光催化活性非常重要。为了排除 BET 的影响,使用 BET 的方法测定物质的比表面积,在 TriStar Ⅱ 3020 V1.03(Micromeritics Instrument 公司,美国)-195.8 ℃ 条件下测试。如图 3.6,从 N_2 的吸脱附等温曲线获得的样品的比表面积分别为 2.7 m^2/g(TiO_2-1)、3.2 m^2/g(TiO_2-2)和 2.5 m^2/g(TiO_2-3),所以 3 个样品的比表面积没有很大差别。

3.2.5
晶面刻蚀蓝色锐钛矿二氧化钛的光电学性质

同样,材料的电子结构也影响着光学性能,因此使用紫外-可见分光光度计(Shimadzu 公司,日本)测试样品的吸收光谱。图 3.7 为 Ti^{3+} 自掺杂的 TiO_{2-x} 微球和对照样的光响应信号。对于 Ti^{3+} 自掺杂的 TiO_{2-x} 微球(图 3.7 中的红线),在可见光区域有大的吸收边,与粉末溶液的颜色变化一致(图 3.7 插图)。这说明非化学计量的 TiO_{2-x} 微球包含大量氧空位。这些结果说明,正如之前的报道,Ti^{3+} 诱导的可见光的吸收能够在 TiO_2 的禁带间形成独立状态[13]。理论研究也表明在块状 TiO_2 中的 Ti^{3+} 和高的氧空位浓度能够在导带下形成电子态空位,有利于减小带隙[25]。

对应的光电特性,在电化学工作站(CH Instruments 公司,美国)上进行电化学测试[30]。暂态光电流和电化学阻抗谱(EIS)在自制的三电极体系中测试,沉积 TiO_2 样品的玻碳电极(GC, an effective area of 0.071 cm^2)作为工作电极,Pt 丝作为对电极,Ag/AgCl 作为参比电极。为说明在电极/电解质界面电荷转移阻力,在开路电压(OCP)下,0.2 mol/L Na_2SO_4 溶液中进行 EIS 谱图的测试和分析,AC 电压振幅 5 mV,频率 10^{-2}~10^5 Hz。图 3.8(a)中样品在紫外光下的光电流响应强度能够反映整个光电子转换过程[31]。样品 TiO_2-2 的光电流分别是 TiO_2-3 和 TiO_2-1 的 2 倍和 6 倍,这说明{101}晶面暴露的 Ti^{3+} 掺杂的 TiO_2 有利于光生电荷的分离和传递。相应的 EIS 结果提供光催化剂中光生电子在界面反映的信息[31]。图 3.8(b)分别为样品 TiO_2-1,TiO_2-2 和 TiO_2-3 光催化剂的 EIS 谱图。TiO_2-2 电极具有更小的弧径,与其他样品相比表现出强化的电化学导电性,小的电子传递阻力,电产生的载体更易分离和转移[32]。

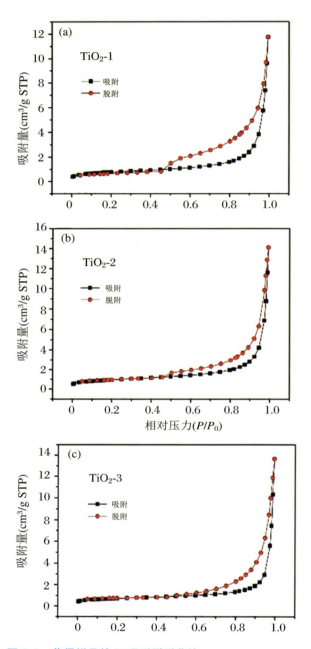

图 3.6 获得样品的 N_2 吸附脱附曲线

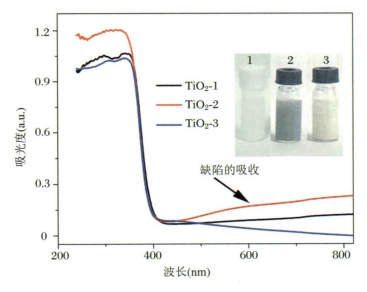

图 3.7 样品的 UV-vis 吸收谱图；内插为相应的样品照片

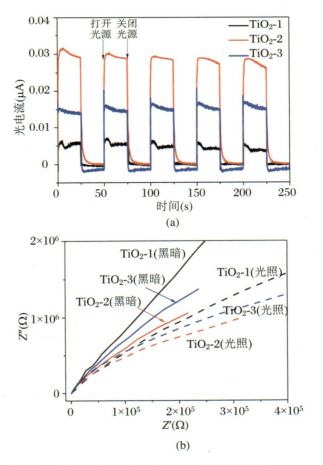

图 3.8 获得样品的光电流对照图(a)和阻抗图(b)

3.3 晶面刻蚀蓝色锐钛矿二氧化钛的光催化产氢性能

为了研究材料的光催化性能,我们使用 TiO_2 催化剂(沉积 1 wt% Pt 作为助催化剂)从水/甲醇(体积比为 4∶1)中光催化产氢实验。光源为 DC 12 V LED (Asahi Spectra 公司,日本)。催化剂(2.0 mg)和 53 μL 的 H_2PtCl_6 溶液(0.02 mg Pt)在磁力搅拌下溶于 1 mL CH_3OH 和 4 mL H_2O 的混合溶液中。通过气相色谱(GC)(Shimadzu 公司,日本)检测 H_2 的产率。图 3.9 为 3 个样品的产氢速率。从一方面,{001} 晶面刻蚀的 TiO_2-3 的产氢速率比 TiO_2-1 高 2.6 倍。{001} 和 {101} 晶面分别是氧化和还原晶面,因此,还原性的 {101} 晶面占主导有利于光催化产氢。另一方面,TiO_2-2 的光催化产氢速率比 TiO_2-3 高 1.3 倍,说明 Ti^{3+} 自掺杂引起的电子结构的变化能够使电子/空穴更有效分离,提高光催化产氢效率。在这种情况下,Ti^{3+} 的自掺杂和晶面刻蚀在电荷分离和电子转移中扮演重要角色[10]。因为 BET 的微小差别,我们也对样品单位表面积的产氢能力进行了探究。归一化的 Pt 负载的样品的产氢速率分别为 116 $\mu mol/(h \cdot m^2)$、

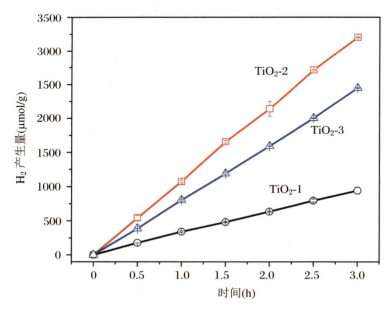

图 3.9　获得样品的产氢实验对照图(70 mW/cm^2,365 nm LED)

334 μmol/(h·m²)和326 μmol/(h·m²)。这个结果也说明晶面刻蚀在光催化中扮演重要角色,而BET表面积和Ti^{3+}掺杂的作用则相对较小。如图3.10所示,在3次循环产氢实验中,Pt/TiO_2-2样品表现出良好的循环稳定性。同理,为了进一步探究材料的稳定性,如图3.11所示,Pt/TiO_2-2样品的EPR图谱与TiO_2-2样品类似,说明TiO_2-2样品在光催化反应后仍然比较稳定,说明了该材料在环境能源方面应用的可能性。

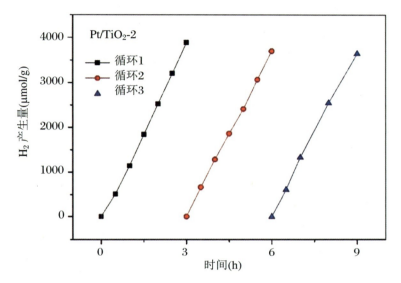

图3.10 Pt/TiO_2-2样品的循环稳定性产氢图(70 mW/cm², 365 nm LED)

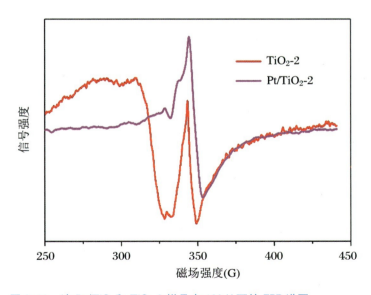

图3.11 该Pt/TiO_2和TiO_2-2样品在130 K下的EPR谱图

3.4 晶面刻蚀蓝色锐钛矿二氧化钛的光催化机制分析

飞秒 TDR 反射光谱,作为一种有效的手段[33],被用来说明在紫外光照射下光催化反应中的载流子动力学。瞬态光谱通过泵浦和探针法,使用由 Nd:YLF 激光器(Spectra-Physics,Empower 15)再生扩增钛蓝宝石激光进行测量(Spectra-Physics,Spitfire Pro F,1kHz,Newport 公司,美国)进行 TDR 测试[34]。钛蓝宝石激光形成脉冲种子(Spectra-Physics,Mai Tai VFSJW;fwhm 80 fs,Newport 公司,美国)。使用光学参量放大器(Spectra-Physics,OPA-800CF-1,Newport 公司,美国)的 4 次谐波(330nm,3μJ pulse-1)作为激发脉冲。在计算机控制的光学延迟之后,白光将通过聚焦在蓝宝石晶体上残余光产生连续脉冲,同时被分为两部分,用作探针和参考光,其中后一个用于补偿激光传递。探针和参考光都被引导到涂在玻璃基底上的样品粉末,反射光通过线性 InGaAs 阵列检测器(Solar,MS3504,Newport 公司,美国)检测。泵浦脉冲被机械斩波器切断,与激光重复率的一半同步,得到一对具有和不具有泵的光谱,由泵浦脉冲引起的吸收变化(吸收率)。所有测量在 25 ℃ 环境温度下进行。

如图 3.12 所示,TiO_2-1 和 TiO_2-3 在 330 nm 的激光照射下产生载流子,电子从 O 1s 的价带激发到 Ti 3d 的导带(表 3.1 中 τ_1 分别为 6.1 ps 和 6.4 ps)。另外,产生的空穴位于 O 1s 的价带上[35]。在表面和块体内的电子达到平衡后(表 3.1 中 τ_2 分别为 41 ps 和 62 ps),在 TiO_2-1 和 TiO_2-3 中产生优先的还原路径,同时电荷会复合(表 3.1 中 τ_3 分别为 471 ps 和 532 ps)。因此更高效率的 TiO_2-3 比 TiO_2-1 表现出更长的寿命(τ_2 = 62 ps;τ_3 = 532 ps)。然而反常的是,更高效率的 TiO_2-2 比 TiO_2-1 和 TiO_2-3 表现出更短的寿命(τ_2 = 33 ps;τ_3 = 342 ps)。这可能是由于 Ti^{3+} 作为能隙态能够捕获载流子[13]。结果是,TiO_2-2(τ_1 = 3.8 ps)在第一衰减阶段短的寿命对应的是电子到达中间带隙 Ti^{3+},然后在表面和体内达到平衡(τ_2 = 33 ps),并且还原反应进行更快(τ_3 = 342 ps)[36]。

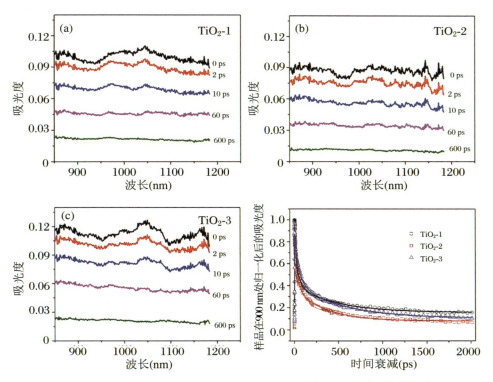

图 3.12 获得样品的 TiO_2-1(a)、TiO_2-2(b) 和 TiO_2-3(c) 的 TDR 光谱以及样品在 900 nm 处归一化后的瞬态吸收谱(d)

表 3.1 紫外激光激发下 TiO_2 催化剂衰减的动力学参数

样品	紫外光(330 nm, 3 μJ/pulse)		
	τ_1(ps)	τ_2(ps)	τ_3(ps)
TiO_2-1	6.1	41	471
TiO_2-2	3.8	33	342
TiO_2-3	6.4	62	532

在本章中，我们利用一步溶剂热法合成了具有{101}晶面的非化学计量的 TiO_{2-x} 微球，并应用于光催化产氢[37]。实验结果表明，非化学计量的高暴露{101}晶面的 TiO_{2-x} 微球表现出最高的光催化性能，分别是煅烧后的样品 TiO_2-3 和非化学计量的 TiO_2-1 的 1.3 和 3.4 倍。这是由于 Ti^{3+} 缺陷能够在 TiO_2 微球的带隙中产生中间带隙，提高电子-空穴分离效率，促进光催化产氢[38]。同时，TiO_2 微球中高暴露{101}还原晶面，有利于电子-空穴在晶面间的分离和传输，进一步促进还原产氢[39]。并且阻抗谱说明材料的本身电阻降低，瞬态吸收光谱则从电子的转移寿命，共同说明材料通过掺杂与还原晶面暴露对光生电子分离起促进作用[40]。该工作提供了简单、有效制备具有高暴露{101}

晶面的非化学计量 TiO_2 的方法，并为设计适用于光催化产氢的 TiO_2 基催化剂提供了新的方向。

参考文献

[1] FUJISHIMA A, HONDA K. Electrochemical photolysis of water at a semi-conductor electrode[J]. Nature, 1972, 238:37-38.

[2] YI M, WANG X, JIA Y, et al. titanium dioxide-based nanomaterials for photocatalytic fuel generations[J]. Chem. Rev., 2014, 114:9987-10043.

[3] LIU G, YANG H G, JIAN P, et al. Titanium dioxide crystals with tailored facets[J]. Chem. Rev., 2014, 114:9559-9612.

[4] LIU L, CHEN X. Titanium dioxide nanomaterials: self-structural modifications[J]. Chem. Rev., 2014, 114:9890-9918.

[5] WANG W K, CHEN J J, ZHANG X, et al. Self-induced synthesis of phase-junction TiO_2 with a tailored rutile to anatase ratio below phase transition temperature[J]. Sci. Rep., 2016, 6:20491.

[6] WALLACE S K, MCKENNA K P. Facet-dependent electron trapping in TiO_2 nanocrystals[J]. J. Phys. Chem. C., 2015, 119:1913-1920.

[7] SELCUK S, SELLONI A. Facet-dependent trapping and dynamics of excess electrons at anatase TiO_2 surfaces and aqueous interfaces[J]. Nat. Mater., 2016, 15:1107.

[8] YU J, LOW J, XIAO W, et al. Enhanced photocatalytic CO_2-Reduction activity of anatase TiO_2 by coexposed {001} and {101} facets[J]. J. Am. Chem. Soc., 2014, 136:8839-8842.

[9] TACHIKAWA T, YAMASHITA S, MAJIMA T. Evidence for crystal-face-dependent TiO_2 photocatalysis from single-molecule imaging and kinetic analysis[J]. J. Am. Chem. Soc., 2011, 133:7197-7204.

[10] LIU X G, DONG G J, LI S P, et al. Direct observation of charge separation on anatase TiO_2 crystals with selectively etched {001} facets[J]. J. Am. Chem. Soc., 2016, 138:2917-2920.

[11] PAN J, LIU G, LU G M, et al. On the true photoreactivity order of {001}, {010}, and {101} facets of anatase TiO_2 crystals[J]. Angew. Chem. Int. Edit., 2011, 50:2133-2137.

[12] WU N Q, WANG J, TAFEN D, et al. Shape-enhanced photocatalytic activity of single-crystalline anatase TiO$_2${101} nanobelts[J]. J. Am. Chem. Soc., 2010, 132:6679-6685.

[13] ZHU Q, PENG Y, LIN L, et al. Stable blue TiO$_{2-x}$ nanoparticles for efficient visible light photocatalysts[J]. J. Mater. Chem. A., 2014, 2:4429-4437.

[14] ZHENG Z K, HUANG B B, MENG X D, et al. Metallic zinc-assisted synthesis of Ti^{3+} self-doped TiO$_2$ with tunable phase composition and visible-light photocatalytic activity[J]. Chem. Commun., 2013, 49:868-870.

[15] NAKAMURA I, NEGISHI N, KUTSUNA S, et al. Role of oxygen vacancy in the plasma-treated TiO$_2$ photocatalyst with visible light activity for NO removal[J]. J. Mol. Catal a-Chem., 2000, 161:205-212.

[16] LEMERCIER T, MARIOT J M, PARENT P, et al. Formation of Ti^{3+} Ions at the surface of laser-irradiated rutile[J]. Appl. Surf. Sci., 1995, 86:382-386.

[17] HASHIMOTO S, TANAKA A. Alteration of Ti 2p XPS spectrum for titanium oxide by low-energy Ar ion bombardment[J]. Surf. Interface Anal., 2002, 34:262-265.

[18] WANG W K, CHEN J J, ZHANG X, et al. Self-induced synthesis of phase-junction TiO$_2$ with a tailored rutile to anatase ratio below phase transition temperature[J]. Sci. Rep., 2016, 6:20491.

[19] WANG W K, CHEN J J, GAO M, et al. Photocatalytic degradation of atrazine by boron-doped TiO$_2$ with a tunable rutile/anatase ratio[J]. Appl. Catal. B-Environ., 2016, 195:69-76.

[20] WANG W K, CHEN J J, LI W W, et al. Synthesis of Pt-loaded self-interspersed anatase TiO$_2$ with a large fraction of {001} facets for efficient photocatalytic nitrobenzene degradation[J]. ACS Appl. Mater. Interfaces., 2015, 7:20349-20359.

[21] FANG W Q, ZHOU J Z, LIU J, et al. Hierarchical structures of single-crystalline anatase TiO$_2$ nanosheets dominated by {001} facets[J]. Chem-Eur. J., 2011, 17:1423-1427.

[22] FAN, ZUO, KRASSIMIR, et al. Active facets on titanium(Ⅲ)-doped TiO$_2$: an effective strategy to improve the visible-Light photocatalytic activity[J]. Angew. Chem. Int. Edit., 2012. DOI: 10.1002/ANIE.

201202191.

[23] LI J-G, R B, M I, TAKAO MORI, et al. Cobalt-doped TiO_2 nanocrystallites: radio-frequency thermal plasma processing, phase structure, and magnetic properties[J]. J Phys Chem C., 2009, 113:8009-8015.

[24] XING M, ZHANG J, FENG C, et al. An economic method to prepare vacuum activated photocatalysts with high photo-activities and photosensitivities[J]. Chem Commun., 2011, 47:4947-4949.

[25] ZUO F, WANG L, WU T, et al. Self-doped Ti^{3+} enhanced photocatalyst for hydrogen production under visible light[J]. J Am Chem Soc., 2010, 132:11856-11857.

[26] CHEN Q, MA W, CHEN C, et al. Anatase TiO_2 mesocrystals enclosed by {001} and {101} facets: Synergistic effects between Ti^{3+} and facets for their photocatalytic performance[J]. Chem-Eur J., 2012, 18:12584-12589.

[27] JI Y, WEI G, CHEN H, et al. Surface Ti^{3+}/Ti^{4+} redox shuttle enhancing photocatalytic H_2 production in ultrathin TiO_2 nanosheets/CdSe quantum dots [J]. J Phys Chem C., 2015, 119. DOI: 10.1021/ACS.JPCC.5b09055.

[28] LIU Y, WANG J, PING Y, et al. Self-modification of TiO_2 one-dimensional nano-materials by Ti^{3+} and oxygen vacancy using Ti_2O_3 as precursor[J]. RSC Adv., 2015, 5. DOI: 10.1039/C5RA07079A.

[29] ZOU X X, LIU J K, SU J, et al. Facile synthesis of thermal- and photo-stable titania with paramagnetic oxygen vacancies for visible-light photocatalysis[J]. Chem-Eur J., 2013, 19:2866-2873.

[30] CHENG X, SHANG Y, CUI Y, et al. Enhanced photoelectrochemical and photocatalytic properties of anatase-TiO_2(B) nanobelts decorated with CdS nanoparticles[J]. Solid State Sci., 2019, 99:106075.

[31] HOU Y, ZUO F, DAGG A, et al. A Three-dimensional branched cobalt-doped alpha-Fe_2O_3 nanorod/$MgFe_2O_4$ heterojunction array as a flexible photoanode for efficient photoelectrochemical water oxidation[J]. Angew. Chem. Int. Ed., 2013, 52:1248-1252.

[32] HOSSEINI Z, TAGHAVINIA N, SHARIFI N, et al. Fabrication of high conductivity TiO_2/Ag fibrous electrode by the electrophoretic deposition method[J]. Phys. Chem. C, 2008, 112(47):18686-18689.

[33] ZHANG P, FUJITSUKA M, MAJIMA T. TiO_2 mesocrystal with nitrogen and fluorine codoping during topochemical transformation: efficient visible

light induced photocatalyst with the codopants[J]. Appl. Catal. B., 2016, 185:181-188.

[34] WANG J, SUN S, DING H, et al. Preparation of a composite photocatalyst with enhanced photocatalytic activity: smaller TiO_2 carried on SiO_2 microsphere[J]. Appl. Surf. Sci., 2019, 493:146-156.

[35] SCHNEIDER J, MATSUOKA M, TAKEUCHI M, et al. Understanding TiO_2 photocatalysis: mechanisms and materials[J]. Chem. Soc. Rev., 2014, 114:9919-9986.

[36] ZHANG P, TACHIKAWA T, FUJITSUKA M, et al. Atomic layer deposition-confined nonstoichiometric TiO_2 nanocrystals with tunneling effects for solar driven hydrogen evolution[J]. J. Phys. Chem. Lett., 2016, 7:1173-1179.

[37] DU Y E, NIU X, LI W, et al. Microwave-assisted synthesis of high-energy faceted TiO_2 nanocrystals derived from exfoliated porous metatitanic acid nanosheets with improved photocatalytic and photovoltaic performance[J]. Materials, 2019, 12. DOI: 10.3390/ma12213614.

[38] MOKLYAK V, CHELYADYN V, HRUBIAK A, et al. Synthesis, structure, optic and photocatalytic properties of anatase/brookite nanocomposites [J]. J. Nano Res., 2020, 64. DOI: 10.4028/www.scientific.net/JNanoR.64.39.

[39] HOAN N, MINH N N, NHI T, et al. TiO_2/diazonium/graphene oxide composites: synthesis and visible-light-driven photocatalytic degradation of methylene blue[J]. J. Nanomater., 2020, 2020:1-15.

[40] SATHIYA N K, BAR-ZIV R, MENDELSON O, et al. Controllable synthesis of TiO_2 nanoparticles and their photocatalytic activity in dye degradation [J]. Mater. Res. Bull., 2020, 126:110842.

第 4 章

{001}二氧化钛介晶结构自生长过程及其光催化性能的研究

4.1 {001}二氧化钛介晶结构自生长研究进展

TiO_2介孔晶体由于其良好的结晶性、多孔性以及有序的定向排列特性获得了极大的关注[1-2]。自组装的TiO_2的催化能力是基于其单个小单位的协同性能和催化潜力。有序自组装的TiO_2介晶存在大量不连续的微观结构,以及具有充分的内部和外部的孔隙率与提高的比表面积。此外,自组装纳米晶体的开放微孔能允许反应物、产物的快速扩散和入射光的良好渗透。纳米晶体结构的排序能够明显影响载流子转移和复合,有序排列的几何体能够促进快速和长距离的载流子转移。另外,TiO_2暴露的{001}晶面上的高密度不饱和的钛原子在催化过程中发挥着重要的作用[6]。由于这些协同效应,TiO_2介晶能被用于光催化反应[7-9]、酶固定化反应[4]、能量储存和转化[3,10-14]。

然而,由于对介晶形成过程的驱动力和机制还缺乏深刻理解,TiO_2介晶目前还很难被制备[1]。尽管拓扑变换是一种有效的合成手段,但很难控制形貌和最终产物的架构。最近有文献报道,TiO_2介晶能通过较为温和的水热、溶剂热或者微乳法合成[3,7-8,10,18]。这些方法中通常使用添加剂来帮助纳米晶体通过各向异性进行成核和生长。另外,考虑到最后的分离和回收,也有TiO_2介晶直接生长在基底上的工作[9,13]。因此,在没有额外添加剂的情况下直接在基底上生长的TiO_2介晶是最直接有效满足催化应用的材料。

在本章中,我们尝试合成一种暴露{001}晶面的锐钛矿六角TiO_2微米棒(Hexagonal Microrod of Anatase TiO_2,HMATT)。该微米棒由介孔的分级TiO_2纳米管通过热处理手段转化而成。这种独特的六角形结构主要由纳米晶体定向自组装而成。该六角形TiO_2不仅具有介孔结构的结构优势,同时还具有单晶结构良好的结晶性、电荷传导能力和高能晶面暴露。

4.2

{001}二氧化钛介晶的合成和表征

分级结构的 TiO_2 纳米管(TNT)作为前驱体是通过电压调控的阳极氧化手段制备的[5]。一个 0.30 mm 钛片浸润在混合溶液(HF：HNO_3：H_2O＝1：1：2, 体积比)中,目的是去除表面的氧化层和污点,接着使用丙酮、异丙醇和去离子水洗净。电解液的成分为 0.09 mol/L NH_4F 溶解在 8 mL 去离子水和 72 mL 乙二醇中。阳极氧化的过程是在双电极体系中进行的,参数为 90 V 持续 2 min,80 V 持续 2 min,70 V 持续 2 min,60 V 持续 2 min,50 V 持续 2 min;接着将这个过程再重复一次,最后 40 V 持续 2 h。然后,将得到的分级 TNT 层不洗涤直接在 500 ℃ 马弗炉中煅烧 3 h,升温速率为 3 ℃/min。最后,在钛片上得到白色 TiO_2 层。

将阳极氧化分级结构 TNT 层作为初始材料的整个制备过程如图 4.1 所示,在反应过程中先形成如图 4.2(a)和(b)所示的分层结构,上层孔直径约为 70 nm,总长约为 0.7 μm,底层孔直径约为 50 nm,长度约为 1.3 μm。这一现象是由于阳极氧化的初始阶段外加短时间的高电压(60~90 V)形成的更为致密的孔径以及后续的中低电压(30~60 V)形成的稍大的孔径。观察到的分层 TNT 的形成很可能是因为在第一层管道中被插入了第二层管道,因此在第一层管道上形成

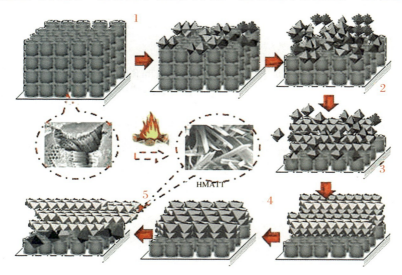

图 4.1　HMATT 制备过程示意图

了一个新的起始层[19]。此外,所制得的上层 TiO_2 纳米管呈破损后的孔状结构,且表面有一层明显的 TiO_2 纳米层覆盖,这层结构可能是源于破损的 TiO_2 纳米管的转化且在高温煅烧过程中成为形成 HMATT 的前驱体。

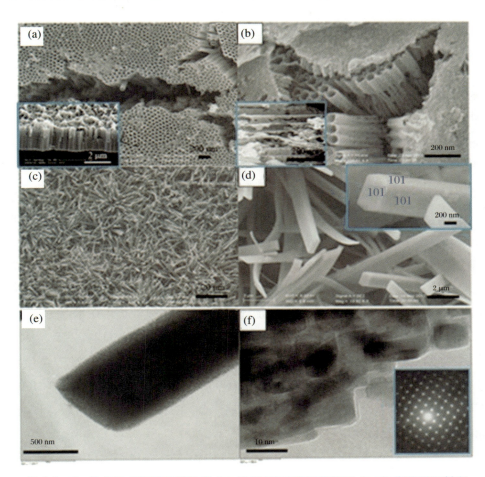

图 4.2　(a、b) TiO_2 分级纳米管前驱;(c、d) HMATT 普通扫描图;(e、f) 高分辨透射和选区电子衍射

4.3 {001}二氧化钛介晶的形成分析

材料的形貌结构、晶格条纹和选区电子衍射分别使用扫描电子显微镜（Sirion 200，FEI 公司，美国）和高分辨透射电子显微镜（JEM-2100，JEOL 公司，日本）获得。在空气下 500 ℃ 煅烧 3 h 之后，上层前驱完全转化为六角形微米棒如图 4.2(c,d)所示。每根六角形微米棒大约有 10 μm 长，1~2 μm 宽，几百 nm 高，暴露出 4 个{101}晶面和 2 个{001}晶面。选区电子衍射显示出"类单晶"的电子衍射。TiO_2 微米棒是由大约平均尺寸为 10 nm 的纳米晶体组成的介孔超结构，如图 4.2(f)所示。纳米晶体之间有 0~5 nm 的孔径。通过对中间过程的表征，发现 HMATT 的形成分为两个过程，如图 4.3 所示。第一个过程是 TiO_2 脚手架形貌的中间体的形成。该形貌主要通过一个溶解和重结晶过程得到。第二个过程是在脚手架形貌的基础上，HMATT 通过一个自组装过程得到。众所周知，在酸性条件下，氟离子(F^-)能加速 TiO_2 的结晶和生长。在本章工作中，上层 TiO_2 的醇解过程是在简单的煅烧条件下通过典型的醇解、成核以及晶体生长系列过程实现的。TiO_2 纳米管被 F^- 不断溶解转化为$[TiF_6]^{2-}$，当 F^- 浓度足够低时，$[TiF_6]^{2-}$ 通过醇解反应转化为 TiO_2：表面吸附的乙二醇（HO—CH_2CH_2—OH）倾向于分离成两个碱氧基（—O—CH_2CH_2—O^-）附着在{001}和{101}晶面。具体反应过程如下：

溶解：$TiO_2 + 6F^- + 4H^+ \longrightarrow [TiF_6]^{2-} + 2H_2O$

重结晶：$\equiv Ti—F + R—OH \longrightarrow \equiv Ti—OR + HF$

$Ti(OR)_n + TiF_n \longrightarrow 2TiO_{n/2} + nRF$

形成的 TiO_2 团簇进一步沿着{001}和{101}方向生长，电极上剩余的乙二醇电解质不仅仅是溶剂热醇解的反应介质，也是一种有利于 TiO_2{001}晶面吸附 F^- 的表面活性剂。同{101}表面比较，{001}晶面上高密度的钛原子有更多选择性将乙二醇以及 F^- 的吸附改变到更理想的状态，这样沿着{001}方向生长的 TiO_2 则生长缓慢。在这一阶段，在乙二醇和 F^- 的协同作用下得到了尺寸约为 10 nm 的暴露{001}晶面 TiO_2 晶体，这些晶体均匀的尺寸和形貌将取代传统固态晶体中原子的角色，定向排列成为具有单元晶胞参数的超结构。因此，我们可以将这些小颗粒 TiO_2 纳米晶体看作组装的原材料，它们沿着{001}方向进行了一个微米尺度的自组装，最终形成了如图 4.3(e)中所示的晶体支架。这一形貌的出现可能是因为乙二醇可能会优先附着在纳米晶体的表面，导致两个支架生长过程中产生强烈的各向异性的相互作用。一旦中间产物脚手架形貌的 TiO_2 形成后，可以作为原始的纳米晶体的自组装支架，最终形成介晶的微米棒。在这一过程中，随着纳米晶体的自组装生长成为六角形微米棒，作为生长支架的 TiO_2 会逐渐以纳米晶体为单位消失，之所以最后这些纳米晶体没有与后来生成的微米棒发生融合的原因可能是两种形貌 TiO_2 具有不同的晶型。

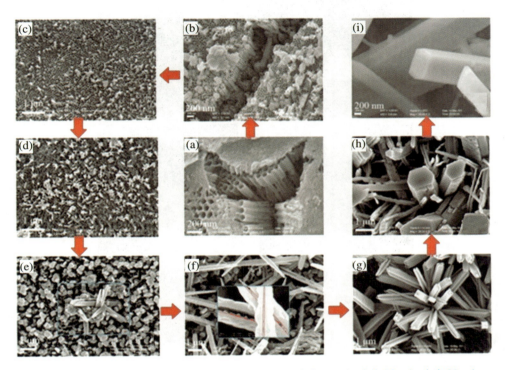

图 4.3　HMATT 生长过程形貌：(a) 0 min；(b) 5 min；(c) 10 min；(d) 30 min；(e) 60 min；(f) 90 min；(g) 120 min；(h) 150 min；(i) 180 min

从图 4.4 可知，在 HMATT 生长过程中产生出的脚手架形貌的中间体同样是由尺寸为 10 nm 作用的 TiO_2 小颗粒单晶定向排列而成。同时脚手架形貌中间的空间会由最初的只有几个小颗粒逐渐通过晶体生长熟化效应被填满生长，最终在中间的空间形成具有特定晶面的六角形截面的 TiO_2 棒状的核。待核生长完毕之后，非稳态的 TiO_2 中间体脚手架的形貌特点会逐渐消失，最后只留下定向排列生长而成的六角形微米棒。

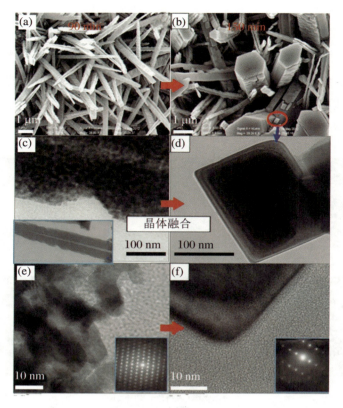

图 4.4　HMATT 生长过程中间体的形貌(SEM\TEM\HRTEM)：(a、c、e)为 90 min;(b、d、f)为 150 min

4.4
{001}二氧化钛介晶的物相分析

材料的物相分析结果通过 X 射线衍射仪获取(Rigaku,日本)。XRD 电流为 200 mA,电压为 40 kV,扫速为 8°/min。材料的比表面积结果通过比表面积和孔隙分析仪获得(Tristar 3020 M,麦克仪器公司,美国)。

如图 4.5 所示,HMATT 的比表面积约为 46.0 m^2/g,同 Degussa P25 数值相似(约为 50.0 m^2/g)。平均孔隙值和平均孔径大约分别为 0.25 cm^3/g 和 7.55 nm。HMATT 的带隙由 DRS 吸收谱推算,如图 4.5(b)所示,其带隙宽度

与纯的 TiO_2 是一致的。图 4.5(c)显示的是 TiO_2 纳米管和 HMATT 的 XRD 图谱。制备得到了分级 TiO_2 纳米管上层结构是非晶的,当在空气下高温煅烧后才会转变为锐钛矿型。所得的 HMATT 衍射峰高度匹配锐钛矿型 TiO_2 标准卡片(JCPDS No. 21-1272)。此外,(004)晶面的衍射峰被检测到比{101}晶面衍射的范围更广,可能意味着纳米晶体的定向生长。同标准峰相比,(004)晶面的峰被明显加强,且一些在(200)晶面和(211)晶面上的峰完全没有。这表明锐钛矿型的小颗粒是对立于基底表面生长的同时主要沿着{001}晶面方向生长[5,23]。总体而言,清晰的 SAED 和 XRD 峰同时表明介晶的微米棒是由含有高比例{001}晶面的纳米小颗粒组成的高孔隙率超结构[24]。

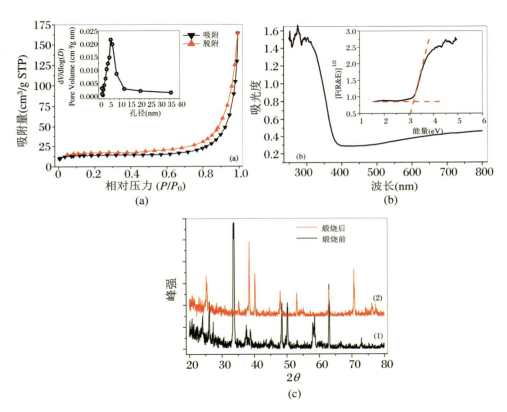

图 4.5 (a) 比表面积测试;(b) DRS 图谱;(c) XRD 表征

4.5 {001}二氧化钛介晶的光催化性能研究

光催化测试是在自制的体积为 60 mL 的石英反应器中进行的。光源为 500 W 氙灯带有波长大于 200 nm 的滤光片。整个光催化降解的过程是在 300～400 nm 的紫外光下进行的。污染物浓度为 50 mL 的双酚 A(BPA,10 mg/L)。每次实验都在同一条件下进行。羟基自由基的测试采用对苯二甲酸捕获,在荧光光谱仪上测试(RF-5301 PC,Shimadzu 公司,日本)。降解后 BPA 的浓度由高效液相检测,波长为 254 nm,VWD(HPLC-1100,Agilent 公司,美国)流动相是水∶甲醇(体积比为 30∶70),流速为 1 mL/min。催化降解的矿化效率通过 TOC 检测仪测得(Vario,Elementar 公司,德国)。

如图 4.6 所示,一方面,同样条件下 HMATT 的荧光强度是 Degussa P25 的 2 倍;另一方面,在荧光强度和光照时间的线性关系中,HMATT 的斜率也大于 P25 的,表明 HMATT 表现出比 P25 更好的光活性。

为了更直接评估 HMATT 的应用潜力,采用了被广泛关注的水中污染物双酚 A(BPA)作为降解对象。如图 4.6(e、f)和图 4.7 所示,在 480 min 的光催化降解实验中,Degussa P25 的去除率只有 46.5%($k_{obs}=0.023$ /h),矿化率仅有 29%。而在同样的时间内,HMATT 已经将 BPA 降解完全($k_{obs}=0.067$ /h),矿化率达到 70%。这些结果表明,HMATT 在双酚 A 降解中表现出持续增强的光催化降解能力。另外,在 5 个循环之内,BPA 的光催化降解效率基本保持稳定,平均降解效率超过 95%。这些结果与光电流的测量结果和羟基自由基测试结果(图 3.5(a、d))是一致的。在以钛片为基底的薄膜催化降解中,介晶 TiO_2 超结构比商业多晶 P25 效果更为突出的原因可以归纳为以下 3 点:① 二维棒状的结构。介晶 TiO_2 在钛片基底上以二维棒状结构存在,这种二维结构有利于电子输运,因此在光催化过程中有利于电子空穴的分离,提高光催化效率。② {001}面的明显暴露。众所周知,{001}面为 TiO_2 的高能晶面,具有很强的导电子能力,在光催化氧化反应中更具有优势[28]。介晶 TiO_2 超结构是由许多单晶 TiO_2 小颗粒沿着{001}方向自组装而成,保留了类似单晶 TiO_2 的导电性和结晶性,同时具有明显暴露的{001}面,不仅有利于电荷输运,降低电子空穴复合,也有利于

BPA 的光催化氧化降解反应。③ 在钛片基底上拥有比 Degussa P25 更大的比表面积。尽管每个 P25 小颗粒具有 50 nm 左右的小尺寸,但是 Degussa P25 在薄片式光催化反应中比表面积并不是很高。而所制备的介晶 TiO_2 超结构在薄膜催化反应中,完好地发挥了微米棒的作用,具有更多的有效催化活性面级,同时由许多单晶 TiO_2 组成的介晶超结构会进一步增大比表面积,为光催化反应提供更多的暴露的活性位点。

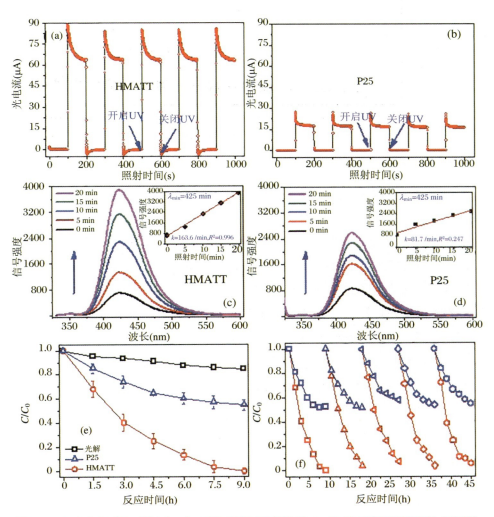

图 4.6 (a,b) 光电流对比;(c,d) 对苯二甲酸荧光强度;(e,f) HMATT 降解双酚 A 和循环稳定性实验

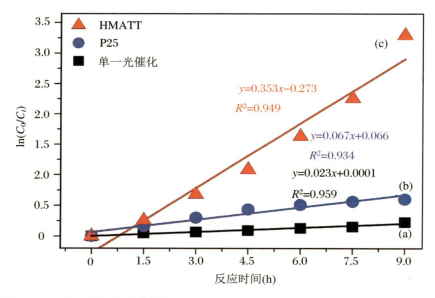

图 4.7 双酚 A 降解动力学曲线

该项研究以钛片为基底，采用分级纳米管前驱体通过煅烧法合成出一种自生长而成的六角形结构介晶 TiO_2 微米棒。同时研究了从分级纳米管前驱体到介晶 TiO_2 微米棒复杂的形成过程。所得的 TiO_2 微米棒是由几纳米大小的 TiO_2 单晶小颗粒有序组装生长而成，整体具有单晶的特点，同时表现出组装材料的结构特点。在钛片基底上，介晶 TiO_2 微米棒具有比单晶材料更高的比表面积，同时能暴露出单晶材料的{001}晶面，与 TiO_2 多晶纳米材料相比具有更为优越的光催化性能。在相同条件下，短电路的光电流强度，光产生的羟基的数量和双酚 A 的降解效率大约分别是 Degussa P25 的 3.2 倍和 2.5 倍。TiO_2 介晶微米棒突出的光催化能力主要归因于其高结晶性，大的孔隙度和暴露出{001}晶面的子单元高度有序的排列。该研究结果为自生长的介孔晶体有序超结构材料的制备提供了新的方法。

参考文献

[1] CÖLFEN H，ANTONIETTI M. Mesocrystals：inorganic superstructures made by highly parallel crystallization and controlled alignment[J]. Angew. Chem. Int. Ed.，2005，44，5576-5582.

[2] ZHANG W, TIAN Y, HE H, et al. Recent advances in the synthesis of hierarchically mesoporous TiO_2 materials for energy and environmental applications[J]. National Science Review，2020，7：1702-1725.

[3] YE J, LIU W, CAI J, et al. Nanoporousanatase TiO$_2$ mesocrystals: additive-free synthesis, remarkable crystalline-phase stability, and improved lithium insertion behaviour[J]. J. Am. Chem. Soc., 2010, 133:933.

[4] CAI J G, QI L M. TiO$_2$ mesocrystals: synthesis, formation mechanisms and applications[J]. Sci. China Chem., 2012, 55:2318-2322.

[5] YI Z, ZENG Y, WU H, et al. Synthesis, surface properties, crystal structure and dye-sensitized solar cell performance of TiO$_2$ nanotube arrays anodized under different parameters[J]. Results Phys, 2019, 15:102609.

[6] YANG H G, SUN C H, QIAO S Z, et al. Anatase TiO$_2$ single crystals with a large percentage of reactive facets[J]. Nature, 2008, 453:638-651.

[7] TARTAJ P. Sub-100 nm TiO$_2$ mesocrystalline assemblies with mesopores: preparation, characterization, enzyme immobilization and photocatalytic properties[J]. Chem. Commun., 2011, 47: 256-260.

[8] TARTAJ P, AMARILLA J M. Multifunctional response of anatase nanostructures based on 25 nm mesocrystal-like porous assemblies[J]. Adv. Mater., 2011, 33:4904-4915.

[9] FILHO E A S, PIERETTI E F, BENTO R T, et al. Effect of nitrogen-doping on the surface chemistry and corrosion stability of TiO$_2$ films[J]. JJ. Mater. Res. Technol, 2020, 9(1): 922-934.

[10] HONG Z, WEI M, LAN T, et al. Additive-free synthesis of unique TiO$_2$ mesocrystals with enhanced lithium-ion intercalation properties[J]. Energy Environ. Sci., 2012, 5:5408-5422.

[11] LOU X W, ARCHER L A. A general route to nonsphericalanatase TiO$_2$ hollow colloids and magnetic multifunctional particles[J]. Adv. Mater., 2008, 20: 1853-1864.

[12] GUO Y G, HU Y S, SIGLE W, et al. Superior electrode performance of nanostructured mesoporous TiO$_2$ (anatase) through efficient hierarchical mixed conducting networks[J]. Adv. Mater., 2007, 19:2087.

[13] CAI J, YE J, CHEN S, et al. Self-cleaning, broadband and quasi-omnidirectional antireflective structures based on mesocrystalline rutile TiO$_2$ nanorod arrays[J]. Energy Environ. Sci., 2012, 5:7575.

[14] LIAO J Y, LEI B X, CHEN H Y, et al. Oriented hierarchical single crystalline anatase TiO$_2$ nanowire arrays on Ti-foil substrate for efficient flexible dye-sensitized solar cells[J]. Energy Environ. Sci., 2012, 5:5750.

[15] ZHOU L, BOYLE D S, O'BRIEN P. A facile synthesis of uniform NH_4TiOF_3 mesocrystals and their conversion to TiO_2 mesocrystals[J]. J. Am Chem. Soc. ,2008,130:1309.

[16] ZHOU L, BOYLE D S, O'BRIEN P. Mesocrystals: a new class of solid materials[J]. Small, 2007,43:144.

[17] FENG J Y, YIN M C, WANG Z Q, et al. Facile synthesis of anatase TiO_2 mesocrystal sheets with dominant {001} facets based on topochemical conversion[J]. CrystEngComm. ,2010,12:3425.

[18] JIN J P, CHEN S T, WANG J M, et al. One-pot hydrothermal preparation of PbO-decorated brookite/anatase TiO_2 composites with remarkably enhanced CO_2 photoreduction activity[J]. Appl. Catal. B: Environ. , 2020, 263:118353.

[19] ROY P, BERGER S, SCHMUKI P. TiO_2 nanotubes: synthesis and applications[J]. Angew. Chem. Int. Ed. ,2011,50:2904.

[20] YANG H Y, YU S F, LAU S P, et al. Direct growth of ZnO nanocrystals onto the surface of porous TiO_2 nanotube arrays for highly efficient and recyclable photocatalysts[J]. Small,2009, 5:2260.

[21] BIAN Z F, TACHIKAWA T, MAJIMA T. Superstructure of TiO_2 crystalline nanoparticles yields effective conduction pathways for photogenerated charges[J]. J. Phys. Chem. Lett. ,2012,3:1422.

[22] YU J C, YU J G, HO W K, et al. Effects of F-doping on the photocatalytic activity and microstructures of nanocrystalline TiO_2 powders[J]. Chem. Mater. , 2002, 14:3808.

[23] LIU B, AYDIL E S. Anatase TiO_2 films with reactive {001} facets on transparent conductive substrate[J]. Chem. Commun. ,2011,47 :9507.

[24] CROSSLAND E J W, NOEL N, SIVARAM V, et al. Mesoporous TiO_2 single crystals delivering enhanced mobility and optoelectronic device performance[J]. Nature, 2013, 495:215.

[25] ZHAO H J, JIANG D L, ZHANG S Q, et al. Photoelectrocatalytic oxidation of organic compounds at nanoporous TiO_2 electrodes in a thin-layer photoelectrochemical cell[J]. J. Catal. ,2007, 250: 102.

[26] LI G Y, LIU X L, ZHANG H M, et al. In situ photoelectrocatalytic generation of bactericide for instant inactivation and rapid decomposition of Gram-negative bacteria[J]. J. Catal. , 2011, 277:88.

[27] LIU G, WANG L Z, SUN C H, et al. Band-to-band visible-light photon excitation and photoactivity induced by homogeneous nitrogen doping in layered titanates[J]. Chem. Mater. ,2009,21:1266.

[28] YU J, LOW J, XIAO W, et al. Enhanced photocatalytic CO_2-reduction activity of anatase TiO_2 by coexposed {001} and {101} facets[J]. J. Am. Chem. Soc. , 2014, 136:8839-8842.

第 5 章

二氧化钛纳米单晶通用元素掺杂改性方法及光催化性能的研究

5.1
二氧化钛纳米单晶掺杂改性研究进展

受限于本身的晶体结构和电子结构,本征态的 TiO_2 具有宽带隙、低电导率、载流子迁移速度缓慢等特点,其光化学和电化学性能不太理想。这些缺点极大地限制了 TiO_2 在光催化、水分解、染料敏化、太阳能电池、锂离子电池等实际体系中的应用[1]。此外,由于 4 价钛离子(Ti^{4+})的典型电子构型,通过向晶格结构中的 s、p、d、f 轨道中掺杂元素来改变 TiO_2 的带隙、电导性和光电形质具有很大的可行性。因此,近年来有许多工作通过掺杂手段调控 TiO_2 晶体结构,解决了 TiO_2 在应用方面的缺陷。由于掺杂提高了 TiO_2 的物化性质,掺杂后的 TiO_2 通常在能源和环境领域均表现出更为突出的性能[2]。例如,把 N 元素引入 TiO_2 的氧的晶格位点中,因为氮和氧的 2p 轨道的杂化从而减小了 TiO_2 的带隙、增强了可见光吸收和光催化活性[3]。

TiO_2 主要通过 3 类元素的掺杂得以改性:非金属元素(S、C、Br、B 等)[3-7]、过渡金属元素(W、Co、V、Sn 等)[8-12]以及稀土金属元素(Sm、Ce、Er、La、Nd 等)[13-17],不同的元素掺杂赋予了 TiO_2 多样的性能。这些掺杂元素可能带来以下 4 种明显的性能改变:① 带隙宽带减小使其能够产生可见光范围吸收,从而在太阳光下获得令人满意的光催化活性(如 N、S、I、B 元素的掺杂)[18];② 增强的电导性带来的更快的电荷迁移使其能有效抑制 TiO_2 表面和体相的无效复合,提高载流子寿命(如 Zn、Fe、Y 元素的掺杂)[19-21];③ 改变 TiO_2 导带价带的位置以及影响载流子转移效率(如 Zr、Nb、W 元素的掺杂);④ 因为电子密度的增强以及更低的电子-空穴复合率导致整体的光化学和电化学活性都得到了全面提升[22-23]。因此,元素掺杂已成为提高 TiO_2 整体催化活性和扩展其催化领域的重要手段之一。

TiO_2 表面和催化属性也和其晶体结构密切相关[24]。{001}晶面暴露的锐钛矿型 TiO_2 单晶因为其高效、廉价、环境友好等优点引起了广泛关注[25]。热力学不稳定的{001}表面已被证明比{101}表面有更高的表面能($0.90\ J/cm^2$)。此外,高能{001}晶面独特的电子结构,表明官能团影响其稳定性、吸附性能和催化活性[26]。所有这些特点使有{001}晶面的 TiO_2 单晶具有更高的能源和环境利用价值。

在 TiO_2 单晶颗粒中,将晶体工程和表面工程有机结合起来,使其应用潜力最大化,具有极高的研究价值[27-33]。然而,实际上很难实现具有明显暴露晶面

的 TiO_2 单晶的掺杂[27]。此外,许多掺杂的前驱体不可避免地会影响 TiO_2 晶体的成核和生长[28]。以前的工作都需要使用一些如 TiN、TiC 和 $TiOF_2$ 等特殊的前驱以及额外的热处理[29-33]。在本章工作中,我们研发了一个简单通用的方法,用以制备不同元素掺杂的具有高暴露{001}晶面的 TiO_2 单晶。Zhang 等曾以废弃的阳极氧化电解液为原料,成功制得了原始的和氮掺杂的 TiO_2 单晶[34],这些晶体在腐殖酸和灭草松的降解中相对于商业 TiO_2(P25)表现出来更高的光催化活性。该工作证实了使用这种电解液作为原料的方法制备的 TiO_2 单晶具有经济和环境双重价值。

在该研究中,我们尝试进一步扩展这种电解废液的用法,发展了一种通用的能满足金属元素、非金属元素和稀土金属元素不同元素掺杂的可行方法;此外,也尝试了双元素掺杂的 TiO_2 单晶制备。我们利用罗丹明染料表征了所制备的掺杂 TiO_2 单晶光催化活性。

5.2 掺杂二氧化钛单晶的合成与表征

通过添加不同的目标元素与阳极氧化的电解液共热,得到掺杂的暴露{001}晶面的 TiO_2 单晶。废弃电解液主要通过上一章工作采用的阳极氧化手段获得,其主要成分为$[TiF_6]^{2-}$,阳极氧化的过程由一个稳压电源控制,以自制双电极体系,2 cm×4 cm 的钛片分别作为阳极和阴极。首先,将两片 0.3 mm 厚的钛片进行煅烧并且浸于混合溶液($HF:HNO_3:H_2O=1:1:2$,体积比)中用于去除表面的氧化层和污点,并且采用丙酮、异丙醇和水洗涤。然后,将 0.09 mol/L NH_4F 溶液(8 mL 去离子水,72 mL 乙二醇)作为电解液。所有电解质都用分析纯试剂制备。电化学处理首先用 80 V 电压电解 2 min;其次,用 70 V 电压电解 2 min,60 V 电压电解 2 min,50 V 电压电解 2 min;最后,在 40 V 电压下保持电解 2 h(温度为室温)。

为了制备 La、Ce、Sn、Fe、B 和 P 掺杂的 TiO_2 晶体,我们采用了 0.05 mL 的硝酸镧(0.1 mol/L)、0.05 mL 的硝酸铈(0.1 mol/L)、5 mL 的氯化亚锡(0.1 mol/L)、0.5 mL 硝酸铁(1 mol/L)、2 mL 硼酸(1 mol/L)和 0.5 mL 磷酸(1

mol/L)分别加入阳极氧化电解液中并超声 15 min。最后,将均匀混合的前驱体溶液转移到合适体积的瓷舟中,升温速率为 3 ℃/min,在 500 ℃或者 600 ℃煅烧 3 h,得到了掺杂产物。

材料的详细形貌和结构使用扫描电子显微镜(Sirion 200,FEI 公司,美国)和球场矫正透射电子显微镜(JEM-ARM200F,JEOL 公司,日本)获得。图 5.1 为暴露{001}晶面的 La、Ce、Sn、Fe、B 和 N 掺杂的 TiO_2 的形貌图。所有掺杂的晶体呈十面体结构,含有 8 个{101}晶面和 2 个{001}晶面。与普通的 TiO_2 单晶对比,这种异质元素的掺杂在一定程度上会影响 TiO_2 单晶的形貌。例如,磷元素掺杂的 TiO_2 的尺寸范围为 50~100 nm,而 La、Ce、Sn、Fe、B 元素掺杂的 TiO_2 的尺寸范围为 200~300 nm,其原因可能是 P 原子的引入影响了 TiO_2 晶体的生长。通过不同样品的形貌对比可以发现,异质原子的引入不会明显改变 TiO_2 的晶体形状,所有掺杂的样品均表现出原始样品的特征。同时 SAED 图谱表明,所得 TiO_2 样品均符合单晶特点。

球差矫正透射电子显微镜的高分辨结果显示,所得 TiO_2 样品具有良好的结晶性和清晰的 0.36 nm 晶格条纹。值得一提的是,因为掺杂量小,所有的掺杂样品和原始样品的晶体表征之间并无显著区别,因此,EDS 映射更进一步佐证掺杂完成。如图 5.1 所示,异质元素均匀分布在掺杂的单晶内部。此外,如磷掺杂样品所示,在 500 ℃煅烧时,所得到的样品为 P、N 共掺杂。

材料的物相分析结果通过 X 射线衍射仪获取(Rigaku,日本)。XRD 电流为 200 mA,电压为 40 kV,扫速为 8°/min。如图 5.2 所示,所得样品的 XRD 图谱全部吻合锐钛矿 TiO_2 标准图谱(JCPDS No.21-1272),且无多余的杂质峰被检出[34]。例如,代表{101}晶面的 25.4°和代表{200}晶面的 48.1°的两个宽峰,以及{004}晶面在 37.93°的峰被检出,表明掺杂的 TiO_2 样品和原始的 TiO_2 样品峰几乎无偏移。但对于某些元素来说,如 La^{3+}(0.115 nm)的尺寸比 Ti^{4+}(0.068 nm)更大[35],由于两种离子半径差异过大,因此,镧掺杂形成的 TiO_2 结构缺陷较少,且掺杂的 TiO_2 单晶几乎不发生变形[36]。材料的拉曼光谱通过在室温下的 325 nm 激光和 Ar^+ 离子激光表征结果(Horiba Scientific 公司,日本)。所得样品的拉曼表征如图 5.2(b)所示,典型的锐钛矿型的拉曼峰 144 cm^{-1}、394 cm^{-1}、514 cm^{-1} 和 638 cm^{-1} 变化甚小。而 147 cm^{-1} 处的拉曼峰发现了不同元素掺杂后明显的红移,在一定程度上证明了 TiO_2 单晶的掺杂。这一红移可归因于结晶程度的提高,掺杂引起的表面氧空位的改变或由于化学状态和离子半径的差异引起的晶体的缺陷。这些改变都有可能减少光电子的复合,因为氧空位和晶体缺陷可以捕获光电子,获得更高的量子效率,从而降低复合率。因此,掺杂有助

于提高单晶 TiO_2 的光催化效率[37]。

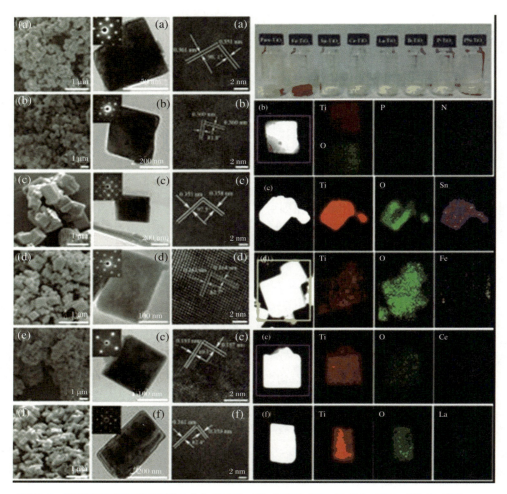

图 5.1 普通形貌和高分辨表征:(a) B-TiO_2;(b) P-TiO_2 SCs;(c) Sn-TiO_2;(d) Fe-TiO_2;
(e) Ce-TiO_2;(f) La-TiO_2

材料的元素分析结果通过 X 射线光电子能谱得到(ESCALAB 250,Thermo-VG Scientific 公司,美国)。通过 XPS 分析可以得到异质元素在 TiO_2 中的含量。图 5.3 表明,掺杂后的钛元素和氧元素的结合能无明显改变。但是,目标元素均表现出了符合掺杂规律的结合能。对于非金属元素 B 来说,在 191.9 eV 处的峰可以归属于 Ti—O—B 键,该结合能位置比纯物相的 B_2O_3 (193.9 eV) 或者 H_3BO_3 (193.5 eV) 更低,这一结果表明硼元素成功掺杂进入 TiO_2 单晶的结构中,而不是以单独的物相存在。此外,钛元素的结合能没有位移的原因可能是其巨大的原子质量。对于非金属元素磷掺杂来说,在 133.3 eV 的 P $2p_{3/2}$ 的结合能表明磷元素呈 P^{5+} 的氧化态,红外结果表明存在形式为 Ti—O—P,证明磷元素掺杂进入了 TiO_2 的晶格内部。有研究表明这种类似的掺杂

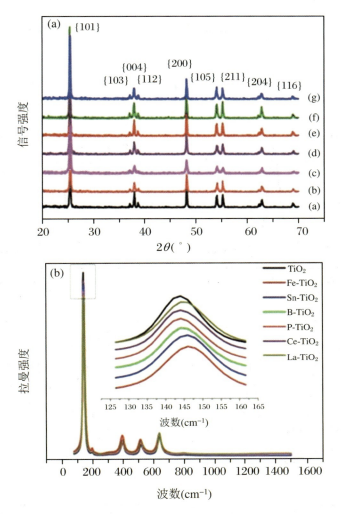

图 5.2 XRD 和 Raman 表征:(a) TiO_2;(b) $La-TiO_2$;(c) $Ce-TiO_2$; (d) $Fe-TiO_2$;(e) $Sn-TiO_2$;(f) $B-TiO_2$;(g) $P-TiO_2$

能使 TiO_2 带隙变窄,从而产生可见光响应[38]。

对于 Sn 掺杂来说,归属为 Sn $3d_{5/2}$(486.6 eV)和 Sn $3d_{3/2}$(494.9 eV)的双峰的出现表明 Sn^{4+} 成功掺入了 TiO_2 晶格,证明 Sn—O—Ti 键的形成。因为 Sn^{4+}(0.69Å m)比 Ti^{4+}(0.53×10Å m)微大,因此在掺杂后,TiO_2 单晶的晶格参数晶胞体积以及 d 值增加[39]。此外,得益于类似的尺寸,Fe^{3+} 的掺杂会相对容易。如图 5.3(d)所示,710.5 eV 和 723.7 eV 两个结合能对应代表Fe_2O_3 中 Fe $2p_{3/2}$ 和 Fe $2p_{1/2}$。这是因为四方晶胞 c 参数的扩展导致铁嵌入钛和氧的空隙之中且保持电荷中立的状态[40]。

对于稀土金属镧元素掺杂来说,只有 La_2O_3 在 TiO_2 中被检测到。如图 5.3(e)所示,836.3 eV 和 853.6 eV 的两个结合能分别归属于 La_2O_3 中 La $3d_{5/2}$

和 La $3d_{3/2}$[41],并且表明没有明显的 Ti—La 键形成。此外,La_2O_3-TiO_2 显示了 O 1s 在 529.0 eV、531.0 eV 和 532.4 eV 三处峰,分别对应于 Ti—O、H—O 的表面能和 La—O 键[42-43]。XPS 结果进一步表明 La_2O_3 在 TiO_2 中存在形势为一直独立的物相。镧改性后的 TiO_2 表现出了显著的光催化性能和热稳定性的提高。

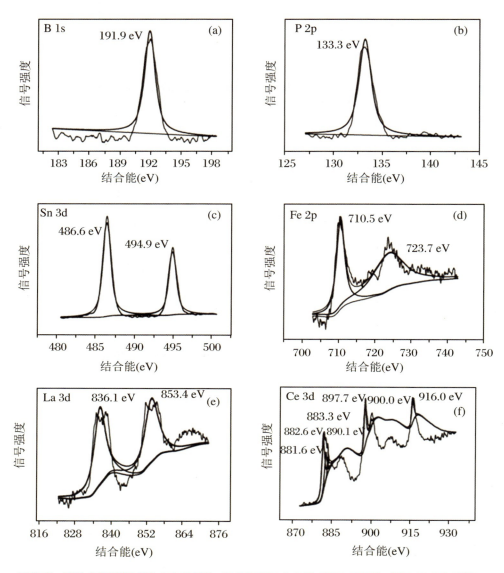

图 5.3　XPS 表征(600 ℃):(a) B-TiO_2;(b) P-TiO_2;(c) Sn-TiO_2;(d) Fe-TiO_2;(e) La-TiO_2;(f) Ce-TiO_2

在稀土金属元素铈掺杂的样品中,882.6 eV、890.1 eV 和 897.7 eV 的 3 个结合能的峰归属于铈掺杂中的 Ce 3d,881.6 eV 则归属于 Ce_2O_3,在 900.0 eV 处的峰表明 Ce^{3+} 有部分被氧化,表现出 Ce_2O_3 的衍射峰。因为从 XRD 图中并

没有看到明显的 Ce_2O_3 的峰，Ce_2O_3 被认为是非晶状态且 Ce—O 键均匀分布在 TiO_2 表面。铈元素的加入不仅影响了晶格结构，同时也影响了晶体生长的尺寸，因为 Ce^{3+}/Ce^{4+} 尺寸（$1.03×10^{-10}$ m/$0.93×10^{-10}$ m）在 Ti^{4+}（$0.68×10^{-10}$ m）和 O^{2-}（$1.32×10^{-10}$ m）之间，所以 Ce 原子可以存在于 TiO_2 晶格中[42-43]。铈掺杂后得到的 TiO_2 单晶尺寸减小的原因可能是前驱体中掺入活泼性更低的铈元素会影响 TiO_2 的结晶温度。当煅烧温度从 600 ℃ 变换至 500 ℃ 时，得到氮元素和外加元素共掺杂的 TiO_2 单晶。在共掺杂的情况下，所得样品的光化学和电化学性质都得到了提高。

5.3 掺杂二氧化钛单晶的电化学性能研究

所有的电化学表征采用三电极系统测试。用于进行电化学测试的玻璃反应器体积约为 50 mL。用一片 6 cm² 的制得的 TiO_2 单晶样品修饰的碳纸或者相同面积的空白碳纸作工作电极，饱和 Ag/AgCl 和铂丝分别用作参比电极和对电极。电化学阻抗谱的测试范围为 $0.01\sim10^5$ Hz，溶液为 10 g/L 的氯化钠溶液，测试环境为开路电压。莫特肖特基曲线测试条件：0.1 mol/L 硫酸钠溶液，频率为 1000 Hz，测试偏压为 $-0.5\sim1.0$ V；OER 反应测试条件：扫速 50 mV/s，0.1 mol/L KOH 溶液。

将所得 TiO_2 单晶均匀沉积正在 FTO 上作为工作电极进行电化学性能测定。图 5.4（a）中 EIS 的图谱显示，与不掺杂的 TiO_2 单晶比较，掺杂的 TiO_2 单晶具有更小的弧，意味着改性后的 TiO_2 有更好的导电性、更小的电阻以及更高的电荷分离效率。从测得的莫特肖特基曲线（图 5.4（b））可知，n 型催化剂的莫特肖特基曲线都存在一个正向的斜率。掺杂的 TiO_2 单晶比不掺杂的 TiO_2 单晶具有更高的斜率，意味着更好的电子传导效率。图 5.4（c、d）是 TiO_2 的 OER 过电位对照图。因为 TiO_2 的过电位被·O^{2-}，HO·和 HOO·等中间产物和 TiO_2 的结合能影响，吸附官能团和催化剂表面强力的结合力意味着更低的过电位。最近有许多文献报道通过掺杂的手段能明显降低 TiO_2 的过电位[2,44]。在我们的工作中，掺杂的 TiO_2 单晶的确比不掺杂的 TiO_2 单晶表现出了更低的 OER 过

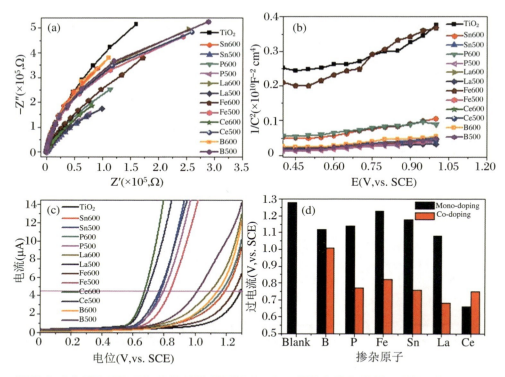

图 5.4 (a) EIS 表征;(b) 莫特-肖特基曲线;(c、d) 不同掺杂样品的产氧过电位对比

电位,表明掺杂能有效改变反应中间产物对催化剂的吸附能力。此外,这样的表面吸附能的变化本质上受掺杂元素的种类和在晶体里掺杂的位置的影响。从图 5.4 可知,在所选择的 0.45 μA 电流下,掺杂的 TiO_2 单晶的过电位顺序为:Ce-TiO_2<La-N-TiO_2<Sn-N-TiO_2<P-N-TiO_2<Fe-N-TiO_2<B-N-TiO_2<La-TiO_2<B-TiO_2<P-TiO_2<Sn-TiO_2<Fe-TiO_2<TiO_2。其中 3% 掺杂的 Ce-TiO_2 表现出最小的(0.66 V)过电位,将不掺杂 TiO_2 的过电位(1.28 V)降低了 40%。

5.4 掺杂二氧化钛单晶的光催化性能研究

材料的光学吸收谱使用紫外-可见分光光度计(Shimadzu 公司,日本)获得。材料的红外光谱通过红外光谱仪(Vertex 70,布鲁克公司,德国)获得。材料的电化学性能通过电化学工作站(CHI 660D,上海辰华公司,中国)。

在试验中考察不同掺杂元素的 TiO_2 单晶样品在可见光条件下对于罗丹明 B 的光降解能力。将 20 mg 的样品分散在 20 ml(5 mg/L)罗丹明 B 溶液中,并置于暗处搅拌 20 min 使催化剂和反应溶液达到吸附解吸附平衡。之后,在光催化实验中,使用 500 W 氙灯搭配 420 nm 的截止滤光片进行辐照(PLS-SXE500,北京泊菲莱公司,中国)。在一定光照时间之后,进行规律性取样。所得样品离心后取上清液进行紫外-可见光谱分析(UV-2450,Shimadzu 公司,日本)。

DRS 图谱表征是检测光催化剂光吸收能力的重要手段。图 5.5 显示所有掺杂的 TiO_2 样品均在 380 nm 处有强吸收(即带隙为 3.20 eV)。然而,掺杂异质

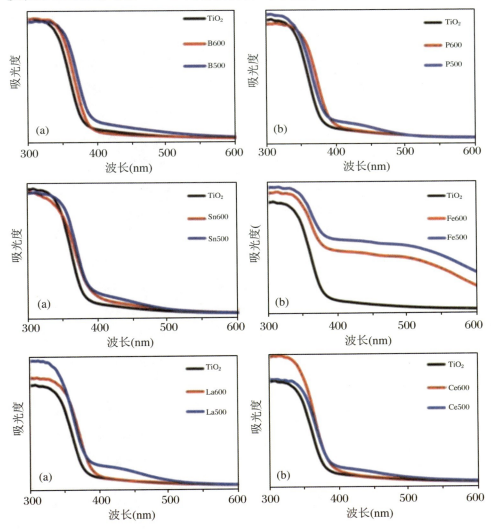

图 5.5 掺杂和不掺杂的 TiO_2 DRS 图谱:(a) B-TiO_2;(b) P-TiO_2;(c) Sn-TiO_2;(d) Fe-TiO_2;(e) La-TiO_2;(f) Ce-TiO_2

元素后，TiO_2 的吸收峰发生红移。这一现象表明掺杂后的 TiO_2 单晶带隙发生了改变。对于 La 和 Ce 掺杂的样品来说，尽管大多数 La_2O_3 和 CeO_2 在 TiO_2 中是一个独立的相，但它们的存在同样能提高比表面积、增强光吸收、捕获光电子和抑制电子-空穴的复合，进一步提高 TiO_2 单晶的光催化性能。另外，铁掺杂后，TiO_2 单晶的吸收峰和颜色变化最明显，呈红棕色（图 5.1）。原则上，光吸收增强是由两个原因引起的：铁离子 3d 轨道上的电子跃迁至 TiO_2 导带上，使吸收带的中心位置升至 400 nm 处以及铁离子的价态变化（$Fe^{3+} + Fe^{3+} \rightarrow Fe^{4+} + Fe^{2+}$）。而对于所有 500 ℃ 掺杂的样品来说，均在 400～500 nm 处检测到了一个明显的肩峰，表明此时的 TiO_2 具有可见光响应。同在 600 ℃ 掺杂的样品比较，此时的吸收带仅在 450 nm 处有一个明显的变化，其余吸收特征都类似。这些现象表明当煅烧温度在 500～600 ℃ 范围时，所得样品能带结构并无明显改变[28,45]。在 500 ℃ 处产生肩峰的主要原因可能是氮元素的掺杂引起晶体缺陷和表面氧空位。这些结果表明，掺杂的 TiO_2 单晶的光电化学特性因为异质元素的掺杂和煅烧得到了不同程度的改进。

罗丹明 B 是一种典型的中染料类污染物，广泛存在于印染废水中。在本章工作中，我们将其作为光催化降解模式污染物用来评价掺杂改性后的 TiO_2 单晶的催化性能，如图 5.6 所示。原始的 TiO_2 单晶因其较宽的带隙在可见光下活性很低，此时测得的罗丹明 B 降解主要源自于罗丹明 B 的染料敏化作用。作为对比，所有掺杂后的 TiO_2 样品由于其晶体和电子结构的改变，表现出很高的罗丹明 B 去除效率（50%～100%）。例如，硼掺杂的 TiO_2 因其中间能带的产生表现出了很好的光催化活性，这是由于硼原子的取代式掺杂促进了大量氧空位的形成[49]。铁掺杂样品的活性提高可以归因于带隙变窄，电子-空穴复合率降低。这些结果进一步表明 TiO_2 单晶成功地被异质元素掺杂改性，且掺杂改性提高了单晶的催化活性。此外，氮元素和其他元素共掺杂的 TiO_2 单晶通常会比单一元素掺杂表现出更好的催化活性。在研究中发现，磷和氮元素共掺杂的 TiO_2 单晶表现出最好的罗丹明 B 去除效率。XPS 结果表明，磷和氮两种元素的存在形式分别为 N—Ti—O 以及 Ti—O—P。根据文献报道，当少量氮元素（1.71%）与磷元素形成 O—P—N 键时，整个 TiO_2 单晶的会表现出优异的催化活性[38]。这一结果也可扩展到其他异质元素。

该研究以废弃的阳极氧化电解液为原料，发展了一种简单通用的进行 {001} 晶面 TiO_2 单晶掺杂的方法，制备了 Ce、La、Sn、Fe、B 和 P 等异质元素掺杂以及 P、N 元素共掺杂的 TiO_2 单晶。同时通过改变煅烧的温度可成功制得单一元素掺杂和异质元素同氮元素共掺杂的 TiO_2 单晶。研究发现异质元素掺杂后的

TiO₂ 单晶并没有发生形貌和结构变化，同样暴露{001}晶面，且表现出来明显增强的可见光催化和电化学反应活性。该工作提供了一个新的制备改性 TiO₂ 单晶的可行方法，为其有效应用于光催化水处理和其他方面提供了新的方案。

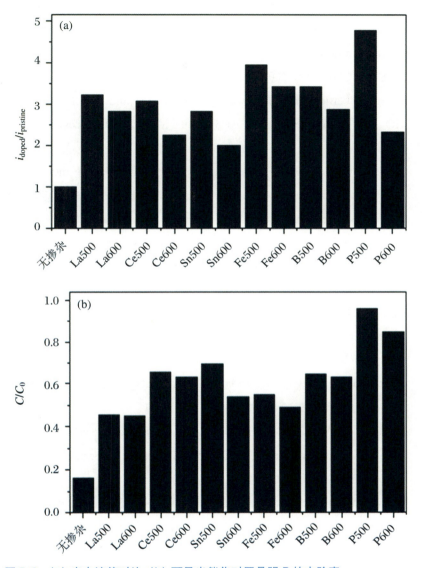

图 5.6 （a）光电流值对比；(b) 可见光催化对罗丹明 B 的去除率

参考文献

［1］ NASR M，EID C，HABCHI R，et al. Recent progress on titanium dioxide nanomaterials for photocatalytic applications[J]. ChemSusChem，2018，11

(18): 3023-3047.

[2] LIU B, CHEN H M, LIU C, et al. Large-scale synthesis of transition-metal-doped TiO$_2$ nanowireswith controllable overpotential[J]. J. Am. Chem. Soc., 2013, 135: 9995-9998.

[3] ASAHI R, MORIKAWA T, OHWAKI T, et al. Visible-light photocatalysis in nitrogen-doped titanium oxides[J]. Science, 2001, 293: 269-271.

[4] XUE X, CHEN H, XIONG Y, et al. Near-infrared-responsive photo-driven nitrogen fixation enabled by oxygen vacancies and sulfur doping in black TiO$_{2-x}$ Sy nanoplatelets[J]. ACS Appl. Mater. Interfaces, 2021, 13(4): 4975-4983.

[5] GU Z, CUI Z, WANG Z, et al. Intrinsic carbon-doping induced synthesis of oxygen vacancies-mediated TiO$_2$ nanocrystals: Enhanced photocatalytic NO removal performance and mechanism[J]. J. Catal., 2021, 393: 179-189.

[6] DU M, QIU B, ZHU Q, et al. Fluorine doped TiO$_2$/mesocellular foams with an efficient photocatalytic activity[J]. Catal. Today, 2019, 327: 340-346.

[7] ABDELRAHEEM W H M, PATIL M K, NADAGOUDA M N, et al. Hydrothermal synthesis of photoactive nitrogen-and boron-codoped TiO$_2$ nanoparticles for the treatment of bisphenol A in wastewater: Synthesis, photocatalytic activity, degradation byproducts and reaction pathways[J]. Appl. Catal. B: Environ., 2019, 241: 598-611.

[8] XING M Y, WU Y M, ZHANG J L, et al. Effect of synergy on the visible light activity of B, N and Fe Co-doped TiO$_2$ for thedegradation of MO[J]. Nanoscale, 2010, 2: 1233-1239.

[9] WANG Y D, CHEN T, MU Q Y. Electrochemical performance of W-doped anatase TiO$_2$ nanoparticles as an electrode material for lithiumion batteries[J]. J. Mater. Chem., 2011, 21: 6006-6013.

[10] MATSUMOTO Y, MURAKAMI M, SHONO T. Room-temperature ferro-magnetism in transparent transition metal doped titanium dioxide[J]. Science, 2001, 291: 854-856.

[11] TSUYUMOTO I, NAWA K. Thermochromism of vanadium-titanium oxide prepared from peroxovanadate and peroxotitanat[J]. J. Mater. Sci., 2008, 43: 985-988.

[12] RADKO M, KOWALCZYK A, MIKRUT P, et al. Catalytic and photocat-

alytic oxidation of diphenyl sulphide to diphenyl sulfoxide over titanium dioxide doped with vanadium, zinc, and tin[J]. RSC Adv., 2020, 10.7: 4023-4031.

[13] RAMAKRISHMA G, DAS A, GHOSH H N. Effect of surface modification on back electron transfer dynamics of dibromo fluorescein sensitized TiO_2 nanoparticles[J]. Langmuir, 2004, 20:1430-1435.

[14] MA T Y, CAO J L, SHAO G S, et al. Hierarchically structured squama-like cerium-doped titania: synthesis, photoactivity and catalytic CO oxidation [J]. J. Phys. Chem. C., 2009, 113:16658-16667.

[15] LIU H, YU L X, CHEN W F, et al. The progress of TiO_2 nanocrystals doped with rare earth ions [J]. J. Nanomater., 2012. DOI: 10.1155/2012/235879.

[16] DAI K, PENG T Y, CHEN H. Photocatalytic degradation of commercial phoxim over La-doped TiO_2 nanoparticles in aqueous suspension[J]. Environ. Sci. Technol.,2009, 43: 1540-1545.

[17] CHEVALLIER L, BAUER A, CAVALIERE S, et al. Mesoporous nanostructured Nb-doped titanium dioxide microsphere catalyst supports for PEM fuel cell electrodes[J]. ACS Appl. Mater. Interfaces,2012, 4: 1752-1759.

[18] LIU G, SUN C H, WANG L Z, et al. Bandgap narrowing of titanium oxidenanosheets: homogeneous doping of molecular iodine for improved photoreactivity[J]. J. Mater. Chem., 2011, 21:14672-14679.

[19] JING L Q, XIN B F, YUAN F L, et al. Effects of surface oxygen vacancies on photophysical and photochemical processes of Zn-doped TiO_2 nanoparticles and their relationships[J]. J. Phys. Chem. B., 2006, 110:17860-17865.

[20] YU J G, XIANG Q J, ZHOU M H. Preparation characterization and visible-light-driven photocatalytic activity of Fe-doped titania nanorods and first-principles study for electronic structures[J]. Appl. Catal. B.,2009, 90: 595-603.

[21] NARAYAN H, ALEMU H, MACHELI L, et al. Synthesis and characterization of Y^{3+}-doped TiO_2 nanocomposites for photocatalytic applications[J]. Nanotechnology, 2009, 20: 255601-255609.

[22] NAGAVENI K, HEGDE M S, MADRAS G. Structure and photocatalytic activity of $Ti_{1-x}M_xO_2^+/$-delta (M=W, V, Ce, Zr, Fe, and Cu) synthesized by solution combustion method [J]. J. Phys. Chem. B., 2004, 108:

20204-20212.

[23] KESSELMAN J M, WERES O, LEWIS N S, et al. Electrochemical production of hydroxyl radical at polycrystalline Nb-doped TiO_2 electrodes and estimation of the partitioning between hydroxyl radical and direct hole oxidation pathways[J]. J. Phys. Chem. B, 1997, 101:2637-2643.

[24] LIU G, YANG H G, PAN J, ET AL. Titanium dioxide crystals with tailored facets[J]. Chem. Rev. ,2014. DOI: 10.1021/cr400621z.

[25] WEI T, NIU B, ZHAO G. Highly characteristic adsorption based on single crystal {001}-TiO_2 surface molecular recognition promotes enhanced oxidation[J]. ACS Appl. Mater. Interfaces,2020,12(35):39273-39281.

[26] PAN J, LIU G, LU G Q, et al. On the true photoreactivity order of {001}, {010}, and {101} facets of anatase TiO_2 crystals[J]. Angew. Chem. Int. Ed. ,2011, 50: 2133-2137.

[27] ZHANG A Y, LONG L L, LI W W, et al. Hexagonal microrods of anatase tetragonal TiO_2: self-directed growth and superior photocatalytic performance[J]. Chem. Commun. ,2013, 49:6075-6077.

[28] LIU G, YANG H G, WANG X W, et al. Visible light responsive nitrogen doped anatase TiO_2 sheets with dominant {001} facets derived from TiN[J]. J. Am. Chem. Soc. , 2009, 131:12868-12869.

[29] XIANG Q J, YU J G, WANG W G, et al. Nitrogen self-doped nanosized TiO_2 sheets with exposed {001} facets for enhanced visible-light photocatalytic activity[J]. Chem. Commun. , 2011, 47:6906-6908.

[30] XIANG Q J, YU J G, JARONIEC M. Nitrogen and sulfur Co-doped TiO_2 nanosheets with exposed {001} facets:synthesis, characterization and visible-light photocatalytic activity[J]. Phys. Chem. Chem. Phys. , 2011, 13: 4853-4861.

[31] ZONG X, XING Z, YU H. Photocatalytic water oxidation on F, N Co-doped TiO_2 with dominant exposed {001}facets under visible light[J]. Chem. Commun. , 2011, 47: 11742-11744.

[32] LUO W Q, FU C Y, LI R F. Er^{3+}-doped anatase TiO_2 nanocrystals:crystal-field levels, excited-state dynamics, upconversion, and defect luminescence [J].Small, 2011, 7:3046-3056.

[33] ROY P, BERGER S, SCHMUKI P. TiO_2 nanotubes:synthesis and applications[J]. Angew. Chem. Int. Ed. ,2011, 50:2904-2939.

[34] ZHANG A Y, LONG L L, YU H Q. Chemical recycling of the wasted anodic electrolyte from TiO₂ nanotubes preparation process to synthesize facet-controlled TiO₂ single crystals as an efficient photocatalyst[J]. Green Chem.,2014,16:2745-2753.

[35] IWASEA M, YAMADAA K, KURISAKIA T, et al. A study on the active sites for visible-light photocatalytic activity ofphosphorus-doped titanium (Ⅳ) oxide particles prepared using aphosphide compound[J]. Appl. Catal.,B.,2013,140:327-332.

[36] WEI J, ZHU W, ZHANG Y Q. Anefficient two-step technique for nitrogen-doped ttanium dioxide synthesizing: visible-lightinduced photodecomposition ofmethylene blue[J].J. Phys. Chem. C.,2007,111:1010-1015.

[37] CHEN D, YANG D, WANG Q, et al. Effects of boron doping on photocatalytic activity and microstructure of titanium dioxide nanoparticles[J]. Ind. Eng. Chem. Res.,2006,45:4110-4116.

[38] LIN L. Synthesis and characterization of phosphor and nitrogen Co-doped titania[J]. Appl. Catal.,B.,2007,76:196-202.

[39] ALDON L, KUBIAK P, PICARD A, et al. Size particle effects on lithium insertion into Sn-doped TiO₂ anatase [J]. Chem. Mater., 2006, 18:1401-1405.

[40] ADÁN C, BAHAMONDE A, FERNÁNDEZ-GARCÍA M, et al. Structure and activity of nanosized iron-doped anatase TiO₂ catalysts for phenol photocatalytic degradation[J].Appl. Catal. B.,2007,72:11-17.

[41] HUO Y N, ZHU J, LI J X, et al. An active La/TiO₂ photocatalyst prepared by ultrasonication-assisted sol-gel method followed by treatment under supercritical conditions[J]. J. Mol. Catal. A: Chem.,2007,278:237-243.

[42] ZHANG J, PENG W Q, CHEN Z H, et al. Effect of cerium doping in the TiO₂ photoanode on the electron transport of dye-sensitized solar cells[J]. J. Phys. Chem. C.,2012,116:19182-19190.

[43] HOU Y, ZUO F, DAGG A, et al. A three-dimensional branched cobalt-doped a-Fe₂O₃ nanorod/MgFe₂O₄ heterojunction array as a flexible photoanode for efficient photoelectrochemical water oxidation[J]. Angew. Chem. Int. Ed.,2013,53:1286-1290.

[44] GARCIA-MOTA M, VOJVODIC A, METIU H, et al. Tailoring the activity for oxygen evolution electrocatalysison rutile TiO₂{110} by transition-metal

substitution[J]. ChemCatChem,2011,3:1607-1611.

[45] LIU G, SUN C H, SMITH S C, et al. Sulfur doped anatase TiO$_2$ single crystals with a high percentage of {001} facets[J]. J. Colloid Interface Sci., 2010, 349:477-483.

[46] HUO Y N, ZHANG X Y, JIN Y, et al. Highly active La$_2$O$_3$/Ti$_{1-x}$B$_x$O$_2$ visible light photocatalysts preparedunder supercritical conditions[J]. Appl. Catal. B,2008,83:78-84.

[47] GUO W, SHEN Y H, BOSCHLOO G, et al. Influence of nitrogen dopants on N-doped TiO$_2$ electrodes and their applications in dye-sensitized solar cells [J]. Electrochim. Acta., 2011, 56:4611-4617.

[48] GUO W, SHEN Y H, WU L Q, et al. Effect of N-dopedamount on the performance of dye-sensitized solar cells based on N-doped TiO$_2$ electrodes[J]. J. Phys. Chem. C., 2011, 115: 21494-21499.

[49] LIU J W, HAN R, WANG H T, et al. Degradation of PCP-Na with La-B co-doped TiO$_2$ series synthesized by the solgel hydrothermal method under visible and solar light irradiation[J]. J. Mol. Catal. A: Chem., 2011, 344: 145-152.

[50] SINGH A P, KUMARI S, SHRIVASTAV R, et al. Satsangi, Iron doped nanostructured TiO$_2$ for photoelectrochemical generation of hydrogen[J]. Int. J. Hydrogen Energy,2008,33:5363-536.

第 6 章

低于相变温度自诱导合成两相比例可控的相结二氧化钛

6.1 相结二氧化钛研究进展

根据文献综述[1-8]，原则上，TiO_2 可以当作完美的催化剂，但是它的有效分离光生电子-空穴的能力还需要持续提高。单独锐钛矿或金红石光生的电子和空穴很容易复合，导致很少电子-空穴能有效达到表面，相应就限制了其光催化活性。

目前为止，普遍认为混相的 TiO_2 有利于减少光生电子-空穴的复合，可以加强光催化反应活性[9]。尤其是商品化 TiO_2——德固赛 P25[10]，作为一种典型的混相 TiO_2，已被广泛地应用于各个领域[11-12]。最近，混相 TiO_2 的相结由于其光催化性能好于单独的纯相，吸引了更多的关注。由于两个物相的导带位置不同，载流子可以从一种物相转移到另外一种物相，这可以延长光催化反应过程中载流子的寿命。这种在两相间有效的载流子分离作为一种有效的方法提高光催化剂活性已经被认知[13-16]。到现在为止，关于相结 TiO_2 合成的一些方法已有报道。例如，通过在前驱体中加入有机试剂或阴离子来合成混相 TiO_2 纳米颗粒[17-19]。混相 TiO_2 薄膜制备则是通过高级等离子电氧化纯的钛片[15]。另外，表面的锐钛矿与金红石相结可以通过对沉积在商品化金红石上的水解钛酸四丁酯制备[16,20]。同样，模板法通过对煅烧的温度来获得物相的转化[21-27]。尽管有一系列的工作是为了制备得到更好的光催化活性 TiO_2 材料，我们迫切需要发展更温和、简单的方法来合成可调控的相结 TiO_2。

在本章中，我们发展一种简单、有效的煅烧方法用以合成锐钛矿与金红石比例可调控的相结 TiO_2，煅烧温度低于相转变温度，且使用 Ti 和 HCl 的水热溶液并无试剂添加。利用罗丹明 B（RhB）的降解和产氢实验评估了所制备相结 TiO_2 的光催化活性。

6.2 两相比例可控的相结二氧化钛的合成与表征

6.2.1 两相比例可控的相结二氧化钛的制备

本章中所使用的试剂均为分析纯[3],且未作任何深度纯化。首先,0.30 mm 厚的钛片需要依次用丙酮、乙醇和去离子水清洗。获得的不同质量钛片(0.6 g、0.4 g、0.2 g、0.1 g 和 0.05 g)依次加到 50 mL 有效容积的高压釜内胆中;依次缓慢加入 20 mL 去离子水和 20 mL 浓盐酸。接着,让样品在高压釜保持在 160 ℃ 条件下 2 h,等自然冷却到室温即可获得无任何沉淀的紫色溶液。紫色溶液通过电子顺磁共振证明 Ti^{3+} 的存在。相对应不同质量钛片所得到的不同溶液用于下一步的合成。

金红石和锐钛矿混相 TiO_2 合成方法如下:在瓷舟内倒入制备的溶液,以 5 ℃/min 的升温速率,在空气中,加热到不同温度(300 ℃、350 ℃、400 ℃、450 ℃、500 ℃、600 ℃、700 ℃ 和 800 ℃)。反应到达上述温度后,快速降温。

同样,制备的不同比例 Ti 片与 HCl 溶液以 5 ℃/min 的升温速率在通空气的马弗炉内煅烧 2 h(500 ℃),接着自然冷却至室温。根据不同钛片与盐酸的比例,获得的样品编号为 $W_1 \sim W_5$。

在准备 Pt 与相结 TiO_2 复合材料时,使用氯铂酸作为前驱体,将 Pt 原位的光还原在金红石/锐钛矿 TiO_2 上;该材料先利用超纯水冲洗 3 遍,接着以 5 ℃/min 的升温速率在通空气的马弗炉中煅烧 0.5 h(450 ℃)。

6.2.2
相结二氧化钛的物相分析及化学性质

根据材料的制备情况,为了确认钛溶液与水和浓盐酸的混合物在衬有特氟隆的不锈钢高压釜中进行水热反应所获得无沉淀的紫色溶液存在 Ti^{3+},通过低温电子顺磁共振光谱(JEOL 公司,日本)在 150 K 下进行测定。利用低温电子顺磁共振(EPR),所得的 EPR 谱,在 $g=1.97$ 处观察到的信号是 Ti^{3+} 离子的顺磁特性,验证了该紫色溶液中 Ti^{3+} 离子的存在(图 6.1)。为了进一步计算 Ti^{3+} 的浓度,通过 TIM865 氧化还原滴定仪(Villeurbanne 公司,法国)测定。氧化还原滴定的原理基于以下氧化还原反应:

$$3Ti^{3+} + Cr(\text{Ⅵ}) \longrightarrow 3Ti^{4+} + Cr(\text{Ⅲ}) \tag{6.1}$$

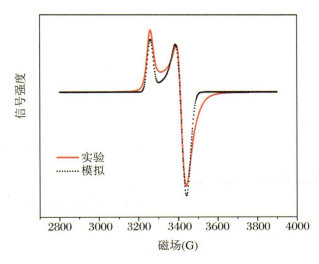

图 6.1 紫色溶液的电子顺磁共振光谱图

Cr(Ⅵ)的消耗量可以在氧化还原滴定过程中获得(图 6.2(a))。从而根据等电点的参数可以计算 Ti^{3+} 离子的浓度。最终通过紫色溶液中的 Ti^{3+} 离子的摩尔量与 Ti 箔的摩尔量对比一致,表明溶液中仅存在作为 Ti 源的 Ti^{3+} 离子。

此外,Ti^{3+} 离子可以与 H_2O 和 Cl^- 配位,形成 $Ti(6H_2O)Cl_3$ 物质作为前驱体,可在后续的合成步骤中进一步应用。可能的形成反应如下:

$$2Ti(foil) + 6H^+ \longrightarrow 2Ti^{3+} + 3H_2 \tag{6.2}$$

$$Ti^{3+} + 6H_2O + 3Cl^- \longrightarrow Ti(6H_2O)Cl_3 \tag{6.3}$$

通过对紫色溶液冷冻干燥,获得的样品进行X射线衍射分析(Rigaku公司,日本),XRD的扫速为8°/min,电压和电流分别为40 kV和200 mA。XRD图谱(图6.2(b))显示紫色溶液由Ti(6H$_2$O)Cl$_3$的组成,与标准卡片(JCPDS No.17-343)一致。

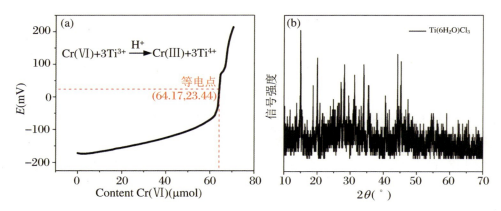

图6.2 (a) 紫色溶液中进行氧化还原滴定测定Ti^{3+}浓度;(b) 冷冻干燥后的紫色溶液样品的XRD图谱

同样,煅烧后样品的物相结果通过X射线衍射分析获得。图6.3(a)显示在不同温度下,煅烧前驱体获得的相结TiO$_2$的XRD。样品的结晶性随着煅烧温度的提高而变好。由于TiO$_2$的光催化活性与结晶性有关,所以所有样品确定煅烧温度为500 ℃。因为通过实验发现高于600 ℃煅烧样品会发生物相的变化。

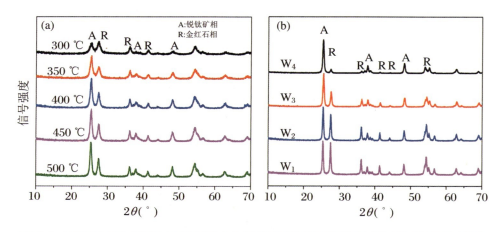

图6.3 相结TiO$_2$纳米颗粒的XRD图:(a) 随温度升高的晶化过程;(b) 以不同Ti片质量获得的样品0.6 g(W_1)、0.4 g(W_2)、0.2 g(W_3)和0.1 g(W_4)

所有样品的XRD都呈现相似的衍射峰,表明所获得的样品均包含锐钛矿和金红石纳米颗粒。图6.3清晰显示了两种TiO$_2$的存在,四方相金红石

(JCPDS No. 21-1276)和四方相锐钛矿(JCPDS No. 21-1272)。由于金红石的(110)衍射峰在 $2\theta = 27.4°$ 处,并且没有任何锐钛矿的峰出现在此位置,所以证明纳米材料中金红石的存在。同理,$2\theta = 25.3°$ 也是锐钛矿的特征峰,证明锐钛矿的存在。综上结果证明金红石和锐钛矿共存在样品 $W_1 \sim W_4$ 中。W_5 号样品则清晰地显示了只有锐钛矿。

根据先前的文献报道[28-29],金红石的质量百分比可以根据个别的衍射峰计算获得。样品的相的含量可以根据以下公式从相应的 XRD 峰的强度计算:

$$f_A = \frac{1}{1 + \frac{1}{K}\frac{I_R}{I_A}}$$
(6.4)
$$K = 0.79, \quad f_A > 0.2$$
$$K = 0.68, \quad f_A \leqslant 0.2$$

其中,f_A 为粉末中锐钛矿相的比例。I_A 和 I_R 则分别表示的是锐钛矿{101}和金红石(110)X 射线衍射峰的强度。

从 $W_1 \sim W_4$ 号样品(图 6.3(b))XRD 可知,在这 4 个混相 TiO_2 样品中,金红石质量百分比分别为 51.8%、46.9%、34.4% 和 11.6%。这个结果显示,以我们的合成方法,可以在很宽的范围内调控金红石与锐钛矿的比例。同时,前驱体的水热时间和温度都不影响样品的混相比例。由于盐酸可以与钛片反应,所以更少计量的钛片可以导致更高 HCl 与 Ti 比例。这样,随着溶液中 HCl 与 Ti 比例的增加,合成样品中金红石含量降低。这个结果符合先前的报道,高 HCl 与 Ti 比例会抑制金红石的形成。

相结 TiO_2 的形成机制可以通过实验结果和第一性原理计算。相对于纯相的锐钛矿或金红石,混相的 TiO_2 由于协同效应[30]和界面电场[31-33],表现更强的光催化活性。金红石和锐钛矿的相结可能存在以下情况:金红石{101}/锐钛矿{001}有相似的晶格排列和参数[31];金红石(100)/锐钛矿(100)可以有完美的重排,即从顶部的锐钛矿层形成金红石层[34],和金红石{111}/锐钛矿{101}拥有很好的晶格匹配[32]。而高分辨透射电子显微镜分析显示了我们可调控相比例的 TiO_2 可能包含以上的相结和一些相对高能的晶体表面的相结,这也导致更倾向于形成界面[35-36]。

两相 TiO_2 的形成反应可能如示意图 6.4 显示[37-39]。为了进一步揭示 TiO_2 样品的内部信息,在室温下分别使用 He-Cd 形成的 325 nm 激光和 Ar^+ 离子 514.5 nm 激光记录紫外-可见拉曼光谱结果(HORIBA Scientific 公司,日本),结果与 XRD 的分析符合[16]。从可见拉曼光谱图 6.5(a)可以观察到两个现象:① 锐钛矿的 E_g 峰在 143 cm^{-1} 处并没有明显的拓宽,说明样品的晶体尺寸在类似的水平；② 除了 W_4 号样处，$W_1 \sim W_3$ 号样金红石都有两个峰在 445 cm^{-1} 处和 612 cm^{-1} 处，表明 W_4 号样的内部几乎没有金红石相。同时，由于紫外拉曼光谱对表面相具有高的敏感性，所以它可以提供关于样品表面相结构的信息[40]。如图 6.5(b)显示，$W_1 \sim W_4$ 号样都能观察到锐钛矿和金红石，确认两相都存在于样品的表面区域。以上的 XRD 和拉曼光谱结果都证实了 W_4 号样的金红石相主要存在于样品表面[41]。

$$2Ti(foil) + 6HCl + 6H_2O \longrightarrow 2TiCl_3 \cdot 6H_2O + 3H_2 \tag{1}$$

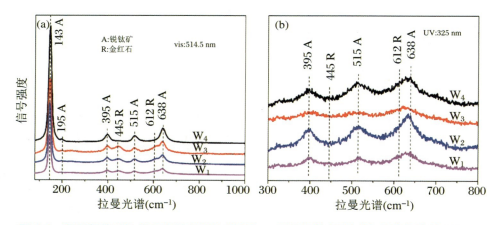

(2)

图 6.4　相结 TiO_2 可能的形成反应机制

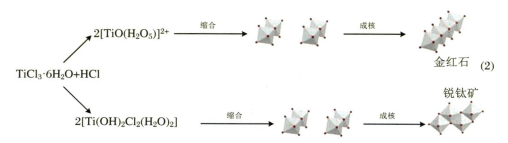

图 6.5　所制得的 TiO_2 样品在激发波长 514.5 nm 可见拉曼光谱(a)和在激发波长 325 nm 紫外拉曼光谱(b)

6.2.3
相结二氧化钛纳米材料的微观结构分析

为了进一步了解样品的微观结构,透射电子显微镜表征(JEM2100,JEOL 公司,日本)被用来观察材料的形貌与结构。图 6.6 显示用不同 Ti^{3+} 浓度合成相结 TiO_2 的低倍透射电子显微镜图。所有的样品呈现了相似的形貌,即大约 40 nm 大小纳米颗粒的团聚体。这个结果也印证了可见拉曼光谱的信息,即所有样品均有一相似的颗粒尺寸。经过 Pt 沉积的纳米颗粒,透射电子显微镜图表明 Pt 纳米颗粒在所有相结 TiO_2 上是均匀的,直径大约为 2 nm。进一步通过电感耦合等离子发射光谱表征,得到 $W_1 \sim W_4$ 号样的 Pt 含量分别为 0.54 mg/mL、0.59 mg/mL、0.53 mg/mL 和 0.55 mg/mL。这样电感耦合等离子发射光谱结果和透射电子显微镜图片共同说明所有样品的 Pt 含量都在相同水平。

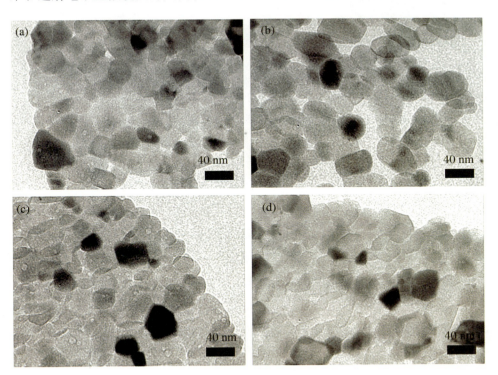

图 6.6 典型的相结 TiO_2 样品的 TEM 图:(a) W_1;(b) W_2;(c) W_3;(d) W_4

为了进一步直观地证明相结的存在,我们获得合成样品的单一颗粒上的高分辨透射电子显微镜图。图 6.7 显示 W_4 号样的高分辨透射电子显微镜图。从图中可以看出,晶格间距为 0.32 nm 对应金红石的(110)面;同时相临近的纳米颗粒晶格间距为 0.35 nm 对应锐钛矿的{101}晶面。由此,高分辨电子显微镜观察证明了锐钛矿与金红石相结的存在。

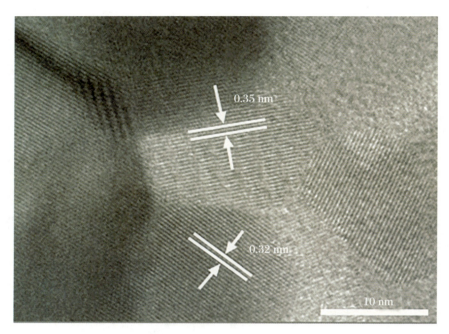

图 6.7　相结 $TiO_2(W_4)$ 样品的 HRTEM 图

6.2.4
相结二氧化钛纳米材料的光物理性质及光化学性质

　　半导体的光学吸收特征是一个很重要的评估它们的光催化性能的因素,这也关联它们的电子带结构[42]。因此,获得材料的光学吸收谱使用紫外-可见分光光度计(Shimadzu 公司,日本)获得,相应的荧光测试使用荧光光度计(Shimadzu公司,日本)。图 6.8(a)紫外-可见吸收光谱显示,所制备的 TiO_2 样品呈现了一个 420 nm 的吸收边。相比较于 $W_1 \sim W_3$ 号样,W_4 号样存在一些蓝移。这个结果可能由于金红石的减少导致光子吸收的范围变小[43]。但是,金红石含量的降低反而提高了材料在小于 400 nm 波长下的吸收。W_4 号样呈现对紫外光的最强吸收。提高紫外光的吸收也可能促进电子空穴的分离[44]。综上,这些结果由于电荷分离,电子从金红石转移到锐钛矿,而空穴向相反的方向转移[45-46]。

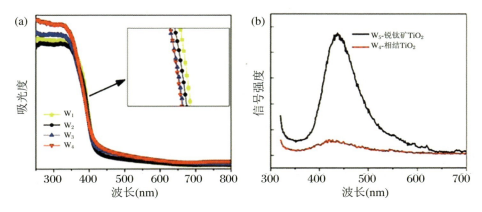

图 6.8 (a) 所有相结 TiO_2 的紫外-可见吸收谱；(b) 锐钛矿和相结 TiO_2 的荧光光谱

另外，相比较于单纯锐钛矿，在波长 300～700 nm 范围内，相结 TiO_2 呈现了更低的光致发光强度（图 6.8(b)）。光致发光是由于光生电子和空穴的复合，相结 TiO_2 的低光致发光强度说明其更低的复合中心密度，存在更多的有效光生载流子用于催化反应[47]。这也说明了相对于其他样品，金红石相的 TiO_2 光生电子和空穴更少。同样，由于存在相结 TiO_2 的光生电子从金红石相到锐钛矿相，从而降低了其电子-空穴复合率[48]。

6.3 相结二氧化钛纳米材料的光催化活性

光催化实验用来评估不同金红石和锐钛矿比例样品的光催化活性。相结 TiO_2 样品的光催化降解 RhB 和产 H_2 都在紫外-可见光下测试。光催化降解 RhB 在 50 mL 烧杯冷水浴中进行。每个烧杯加入 20 mg TiO_2 催化剂，分散在 30 mL RhB 溶液中。RhB 的吸收光谱测试则通过紫外-可见分光光度计（UV-2401PC，Shimadzu 公司，日本）记录。在密封的石英反应池，Pt/TiO_2 分散在甲醇水溶液中进行光催化产氢。光源为 350 W 氙灯（CHF-XM-350W，Beijing Trusttech 公司，中国）。0.1% 的 Pt 负载的 TiO_2 分散在 10 mL 甲醇和 90 mL 水溶液里，并且搅拌。所产生的 H_2 利用气相色谱（SP-6890，Lunan 公司，中国）检测，该色谱配备有 TCD 检测器，以 Ar 气作为载气。

图 6.9(a)显示 TiO$_2$ 纳米材料光催化降解 RhB。空白组实验(无催化剂)证明 RhB 在紫外-可见光照下的稳定性。相比较,TiO$_2$ 的加入提高了 RhB 的降解。如图 6.9(a)显示,相结的光催化活性明显好于单独的锐钛矿(W$_5$)和金红石(W$_6$)。经过 90 min 照射,W$_4$ 号样对 RhB 展现最高的光催化降解效率(92.4%)。则其他样品 W$_1$、W$_2$ 和 W$_3$ 对 RhB 的降解效率分别为 76.4%、81.0% 和 87.4%。对于所有样品,RhB 的光催化降解效率随着样品中金红石含量的降低而明显提高。这些结果表明金红石与锐钛矿比例和表面的相结在控制混相 TiO$_2$ 光生电子-空穴复合扮演重要作用[49],相对应影响其光催化活性。从紫外-可见拉曼光谱看,W$_4$ 号样的内部更多的为锐钛矿,而表面存在少量的金红石,在 TiO$_2$ 表面暴露更多的相结。这样有利于在 TiO$_2$ 表面吸附的氧和水分子更好捕获光生电子-空穴形成活性氧化物[16]。

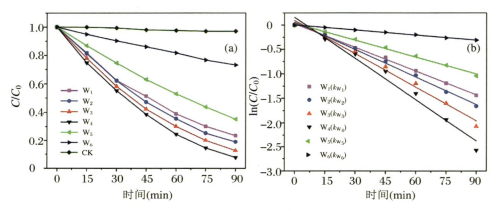

图 6.9 在 UV-vis 照射下,利用制备的 TiO$_2$ 光催化降解溶液中的 RhB
C_0 和 C 分别代表 $t=0$ 和 $t=t$ 时的 RhB 浓度

在图 6.9(b)中,RhB 光催化降解曲线呈现遵循一级动力学。这样,RhB 降解速率常数(k)可以通过 $\ln(C/C_0)$ 与时间的拟合直线斜率得出,C 代表 RhB 的浓度。表 6.1 显示了各个光催化剂对 RhB 降解的速率常数对比,相结的 TiO$_2$ 对 RhB 进行了最快的降解。同时,k_{W_4} 大于 $k_{W_1} \sim k_{W_3}$ 和 $k_{W_5} \sim k_{W_6}$,说明在一定范围内,相结 TiO$_2$ 中低比例的金红石更有利于加强光催化能力。

表 6.1　不同金红石含量的 RhB 降解速率常数(k)和线性相关系数(R^2)

催化剂编号	金红石含量（wt%）	k(/min) 速率常数	R^2 线性相关系数
W_1	51.8	−0.016	0.998
W_2	46.9	−0.018	0.996
W_3	34.4	−0.022	0.987
W_4	11.6	−0.028	0.973
W_5	0	−0.012	0.994
W_6	100	−0.004	0.998

另外，光催化产氢实验也用来评估不同相比例的相结 TiO_2（沉积的 Pt 作为共催化剂）。这些样品是通过对不同浓度的前驱体在相同温度下煅烧获得。图 6.10 显示了随着金红石与锐钛矿比例变化，相结 TiO_2 产氢速度变化；相结 TiO_2 的产氢速度远远大于单独的锐钛矿（W_5）和金红石（W_6）。尤其，整个光催化活性与催化剂表面的金红石含量相关。从 W_1～W_4 号样观察，氢气的产生速率随着金红石含量降低而升高。同样，产氢实验也用来研究无 Pt 纳米颗粒负载的相结 TiO_2 的活性。如图 6.10(b) 显示，最开始产生的 H_2 比较少是由于最初的光生电子用于形成氧空位。在 1 h 后，氢气产生速率趋于稳定，这 4 种相结 TiO_2 纳米颗粒（无 Pt）呈现了相似的光催化效率趋势。值得注意的是，根据紫外-可见拉曼光谱分析，金红石与锐钛矿的比例变化不会改变材料的表面相和内部相。该结果暗示，表面相结内部主要为锐钛矿和表面少量金红石的形式，更加有利于氢气产生。

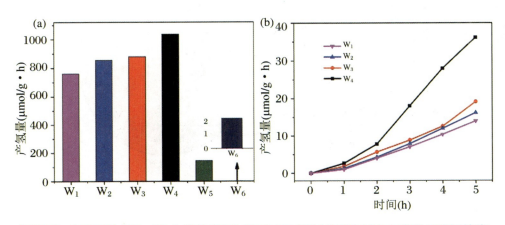

图 6.10　相结 TiO_2（W_1～W_4）、锐钛矿（W_5）和金红石（W_6）分别在 UV-vis 照射下产氢性能图：(a) 存在 Pt 助催化剂；(b) 无 Pt 助催化剂

6.4 相结二氧化钛纳米材料的光催化活性机制研究

以上的结果揭示,相结 TiO_2 代表了高光催化活性。在锐钛矿和金红石间形成的表面相结加强了颗粒间有效的电子传递,促进更多电荷分离[16,47,50]。同时,含少量金红石的两相混合的 TiO_2 呈现更高的光催化活性[51]。相反,更高金红石含量样品的光催化活性降低,这是由于表面锐钛矿 TiO_2 可能全被金红石覆盖,导致暴露的相结减少。

基于以上分析,关于相结 TiO_2 在紫外-可见光下的光催化机制推测如图 6.11 所示。对于相结样品,电子从金红石移动到锐钛矿,而空穴则反之[52]。这样,它就可以促进光激发的电子空穴对的分离[43,53]。所以更好的 TiO_2 光催化活性主要是由于两相间的协同效应。

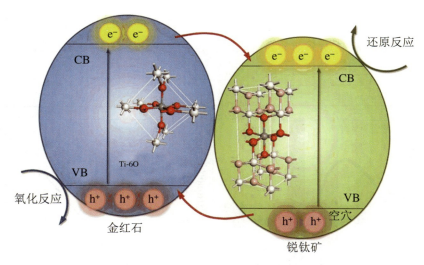

图 6.11 锐钛矿和金红石界面的一种可能导带价带排列形式
红色箭头分别表示电子、空穴导带价带的流动。e^- 和 h^+ 分别代表电子和空穴

在光催化降解 RhB 过程,·OH 作为一种主导降解的活性物质[47]。氧气可以捕获激发并转移到 TiO_2 表面的电子形成 O_2、HO_2 和 H_2O_2,接着 O_2 或电子进

一步与 H_2O_2 反应生成·OH。而在 TiO_2 表面的光生空穴与 OH^- 和 H_2O 反应产生·OH。·OH 则被认为是主要的反应活性物种，可以降解污染物[54]。由此，不仅光生电子-空穴对被有效地利用，而且电子-空穴的复合（荧光发射过程）得到了抑制。同时，它也显示了光生电子-空穴在半导体表面转移并形成反应活性物种的机制。

类似，光催化产 H_2 过程中，在 Pt 负载的 TiO_2 光催化剂上激发的电子转移到 Pt 纳米颗粒。这就加强了 Pt 与 TiO_2 的作用和界面电子的转移和分离，可以更高效地分解水来产 H_2[55]。因此，相结的 TiO_2 对光氧化和光还原有更高紫外-可见光催化活性。对此 TiO_2 复合材料，金红石与锐钛矿比例提高了界面电子传递的有效性和光催化活性。

在本章中，我们发展了一种简单且有效的方法用于制备金红石与锐钛矿比例可控的相结 TiO_2 纳米材料。该方法既不需要调节剂来调控相的比例，也无需温度调节来使相发生转变。实验结果表明，前驱物的浓度可以调控相结 TiO_2 的合成，且金红石含量随着 Ti 源浓度的降低而减少；金红石和锐钛矿相结 TiO_2 的光催化活性随金红石含量的降低而提高，且均高于单独的金红石相或锐钛矿相 TiO_2；与其他催化剂相比，金红石 11.6% 的 TiO_2 对 RhB 展现最高的光催化降解效率（92.4%）和产氢能力（大于 1 mmol/(g·h)），这是因为颗粒内部的低活性金红石含量减少，而表面的金红石锐钛矿相结形成，促进了光生载流子分离。从而证明金红石与锐钛矿比例合适的相结 TiO_2 可以促进表面两相间的电子传递。该工作提供一种制备相结的方法，且加深对混相 TiO_2 在光催化、光电化学和光电领域应用的了解。

参考文献

[1] FUJISHIMA A, HONDA K. Photolysis-decomposition of water at the surface of an irradiated semiconductor[J]. Nature, 1972, 238:37-38.

[2] ITO S, ZAKEERUDDIN S, HUMPHRY-BAKER R. High-efficiency organic-dye-sensitized solar cells controlled by nanocrystalline-TiO_2 electrode thickness[J]. Adv. Mater., 2006, 18:1202-1205.

[3] LI G, PARK S, KANG D W, et al. 2,4,5-Trichlorophenol degradation using a novel TiO_2-coated biofilm carrier: Roles of adsorption, photocatalysis, and biodegradation[J]. Environ. Sci. Technol., 2011, 45:8359-8367.

[4] CHEN C, MA W, ZHAO J J C S R. Semiconductor-mediated photodegra-

dation of pollutants under visible-light irradiation[J]. Chem. Soc. Rev., 2010, 39:4206-4219.

[5] YANG Z, DU G, MENG Q, et al. Synthesis of uniform TiO$_2$@carbon composite nanofibers as anode for lithium ion batteries with enhanced electrochemical performance[J]. J. Mater. Chem., 2012, 22:5848-5854.

[6] YI M, WANG X, JIA Y, et al. Titanium dioxide-based nanomaterials for photocatalytic fuel generations[J]. Chem. Rev., 2014, 114:9987-10043.

[7] BAI Y, MORA-SERó I, ANGELIS F D, et al. Titanium dioxide nanomaterials for photovoltaic applications[J]. Chem. Rev., 2014, 114:10095-10130.

[8] LIU G, YANG H G, JIAN P, et al. Titanium dioxide crystals with tailored facets[J]. Chem. Rev., 2014, 114:9559-9612.

[9] MANIKANDAN B, MURALI K R, JOHN R. Optical morphological and microstructural investigation of TiO$_2$ nanoparticles for photocatalytic application[J]. Iranian J. Catal., 2021, 11:1-11.

[10] OHNO T, SARUKAWA K, TOKIEDA K, et al. Morphology of a TiO$_2$ photocatalyst (Degussa, P25) consisting of anatase and rutile crystalline phases- science direct[J]. J. Catal., 2001, 203:82-86.

[11] HURUM D C, AGRIOS A G, GRAY K A, et al. Explaining the enhanced photocatalytic activity of Degussa P25 mixed-phase TiO$_2$ using EPR[J]. J. Phys. Chem. B, 2003, 107:4545-4549.

[12] BICKLEY R I, GONZALEZ-CARRENO T, LEES J S, et al. A structural investigation of titanium dioxide photocatalysts[J]. J. Solid. State. Chem., 1991, 92:178-190.

[13] ANGELIS F D, VALENTIN C D, FANTACCI S, et al. Theoretical studies on anatase and less common TiO$_2$ phases: Bulk, surfaces, and nanomaterials [J]. Chem. Rev., 2014, 114:9708-9753.

[14] LIAO Y, QUE W, JIA Q, et al. Controllable synthesis of brookite/anatase/rutile TiO$_2$ nanocomposites and single-crystalline rutile nanorods array[J]. J. Mater. Chem., 2012, 22:7937-7944.

[15] SU R, BECHSTEIN R, S L, et al. How the anatase-to-rutile ratio influences the photoreactivity of TiO$_2$[J]. J. Phys. Chem. C, 2011, 115:24287-24292.

[16] ZHANG J, XU Q, FENG Z, et al. Importance of the relationship between

surface phases and photocatalytic activity of TiO_2[J]. Angew. Chem. Int. Ed., 2008, 47:1766-1769.

[17] ZHAO B, LIN L, HE D J J O M C A. Phase and morphological transitions of titania/titanate nanostructures from an acid to an alkali hydrothermal environment[J]. J. Mater. Chem. A, 2013, 1:1659-1668.

[18] LI G, CISTON S, SAPONJIC Z V, et al. Synthesizing mixed-phase TiO_2 nanocomposites using a hydrothermal method for photo-oxidation and photo-reduction applications[J]. J. Catal., 2008, 253:105-110.

[19] YAN M, CHEN F, ZHANG J, et al. Preparation of controllable crystalline titania and study on the photocatalytic properties[J]. J. Phys. Chem. B, 2005, 109:8673-8678.

[20] KAWAHARA T, OZAWA T, IWASAKI M, et al. Photocatalytic activity of rutile-anatase coupled TiO_2 particles prepared by a dissolution-reprecipitation method[J]. J. Colloid Interf. Sci., 2003, 267:377-381.

[21] GANG X, RUI S, DROUBAY T C, et al. Photoemission electron microscopy of TiO_2 anatase films embedded with rutile nanocrystals[J]. Adv. Funct. Mater., 2007, 17:2133-2138.

[22] GANG L, YAN X, CHEN Z, et al. Synthesis of rutile-anatase core-shell structured TiO_2 for photocatalysis[J]. J. Mater. Chem., 2009, 19:6590-6596.

[23] HUANG W, TANG X, WANG Y, et al. Selective synthesis of anatase and rutile via ultrasound irradiation[J]. Chem. Commun., 2000:1415-1416.

[24] LI S, CHEN J, ZHENG F, et al. Synthesis of the double-shell anatase-rutile TiO_2 hollow spheres with enhanced photocatalytic activity[J]. Nanoscale, 2013, 5:12150-12155.

[25] PAN L, HUANG H, LIM C K, et al. TiO_2 rutile-anatase core-shell nanorod and nanotube arrays for photocatalytic applications[J]. RSC Adv., 2013, 3:3566-3571.

[26] LIU B, KHARE A, AYDIL E S J A A M, et al. TiO_2-B/anatase core-shell heterojunction nanowires for photocatalysis[J]. ACS Appl. Mater. Interfaces, 2011, 3:4444-4450.

[27] WU M, LIN G, CHEN D, et al. Sol-hydrothermal synthesis and hydrothermally structural evolution of nanocrystal titanium dioxide[J]. Chem. Mater., 2002, 14:1974-1980.

[28] SHAN C, XU Y J T J O P C C. Explaining the high photocatalytic activity of a mixed phase TiO_2: a combined effect of O_2 and crystallinity[J]. J. Phys. Chem. C, 2011, 115:21161-21168.

[29] LI G, GRAY K A J C O M. Preparation of mixed-phase titanium dioxide nanocomposites via solvothermal processing[J]. Chem. Mater., 2007, 19: 1143-1146.

[30] KHO Y K, IWASE A, TEOH W Y, et al. Photocatalytic H_2 evolution over TiO_2 nanoparticles: the synergistic effect of anatase and rutile[J]. J. Phys. Chem. C, 2010, 114:2821-2829.

[31] XIA T, LI N, ZHANG Y, et al. Directional heat dissipation across the interface in anatase-rutile nanocomposites[J]. ACS Appl. Mater. Interfaces, 2013, 5:9883-9890.

[32] JU M G, SUN G, WANG J, et al. Origin of high photocatalytic properties in the mixed-phase TiO_2: a first-principles theoretical study[J]. ACS Appl. Mater. Interfaces, 2014, 6:12885-12892.

[33] ZHANG X, LIN Y, HE D, et al. Interface junction at anatase/rutile in mixed-phase TiO_2: formation and photo-generated charge carriers properties [J]. Chem. Phys. Lett., 2011, 504:71-75.

[34] DESKINS N A, KERISIT S, ROSSO K M, et al. Molecular dynamics characterization of rutile-anatase interfaces[J]. J. Phys. Chem. C, 2007, 111:9290-9298.

[35] GONG X Q, SELLONI A, BATZILL M, et al. Structure and energetics of step edges on anatase TiO_2{101}[J]. Nat. Mater., 2006, 5:665-670.

[36] YANG H G, SUN C H, QIAO S Z, et al. Anatase TiO_2 single crystals with a large percentage of reactive facets[J]. Nature, 2008, 453:638-642.

[37] ZHENG W, LIU X, YAN Z, et al. Ionic liquid-assisted synthesis of large-scale TiO_2 nanoparticles with controllable phase by hydrolysis of $TiCl_4$[J]. ACS Nano., 2009, 3:115-122.

[38] HU X, LI W, ZHENG Y, et al. Influence of solution concentration on the hydrothermal preparation of titania crystallites[J]. J. Mater. Chem., 2001, 11:1547-1551.

[39] POTTIER A, CHANéAC C, TRONC E, et al. Synthesis of brookite TiO_2 nanoparticles by thermolysis of $TiCl_4$ in strongly acidic aqueous media[J]. J. Mater. Chem., 2001, 11:1116-1121.

[40] CIALLA-MAY D, SCHMITT M, POPP J. Theoretical principles of Raman spectroscopy[J]. Phys. Sci. Rev., 2019, 4:9.

[41] ZHANG J, LI M, FENG Z, et al. UV Raman spectroscopic study on TiO_2. I. Phase transformation at the surface and in the bulk[J]. J. Phys. Chem. B, 2006, 110:927-935.

[42] WANG X T, GUO L. SERS activity of semiconductors: crystalline and amorphous nanomaterials[J]. Angew. Chem. Int. Ed., 2020, 59: 4231-4239.

[43] SCANLON D O, DUNNILL C W, BUCKERIDGE J, et al. Band alignment of rutile and anatase TiO_2[J]. Nat. Mater., 2013, 12:798-801.

[44] MORIYAMA A, YAMADA I, TAKAHASHI J, et al. Oxidative stress caused by TiO_2 nanoparticles under UV irradiation is due to UV irradiation not through nanoparticles[J]. Chem. Biol. Interact., 2018, 294:144-150.

[45] HUNG W C, CHEN Y C, CHU H, et al. Synthesis and characterization of TiO_2 and Fe/TiO_2 nanoparticles and their performance for photocatalytic degradation of 1, 2-dichloroethane[J]. Appl. Surf. Sci., 2008, 255: 2205-2213.

[46] BIAN Z, TACHIKAWA T, KIM W, et al. Superior electron transport and photocatalytic abilities of metal-nanoparticle-loaded TiO_2 superstructures[J]. J. Phys. Chem. C, 2012, 116:25444-25453.

[47] YANG L, LUO S, YUE L I, et al. High efficient photocatalytic degradation of p-nitrophenol on a unique Cu_2O/TiO_2 p-n heterojunction network catalyst[J]. Environ. Sci. Technol., 2010, 44:7641-7646.

[48] YUN T K, PARK S S, KIM D, et al. Effect of the rutile content on the photovoltaic performance of the dye-sensitized solar cells composed of mixed-phase TiO_2 photoelectrodes[J]. Dalton Trans., 2012, 41:1284-1288.

[49] ETACHERI V, SEERY M K, HINDER S J, et al. Highly visible light active $TiO_{2-x}N_x$ heterojunction photocatalysts[J]. Chem. Mater., 2010, 22: 3843-3853.

[50] KIM W, TACHIKAWA T, MOON G, et al. Molecular-level understanding of the photocatalytic activity difference between anatase and rutile nanoparticles[J]. Angew. Chem. Int. Ed., 2014, 53:14036-14041.

[51] LIU Y A, ZOU X H, LI L F, et al. Engineering of anatase/rutile TiO_2 heterophase junction via in-situ phase transformation for enhanced photocat-

alytic hydrogen evolution[J]. J. Colloid Interf. Sci., 2021, 599:795-804.

[52] HU K K, E L, ZHAO D, et al. Characteristics and performance of rutile/anatase/brookite TiO_2 and TiO_2-$Ti_2O_3(H_2O)_2(C_2O_4)·H_2O$ multiphase mixed crystal for the catalytic degradation of emerging contaminants[J]. CrystEngComm, 2020, 22:1086-1095.

[53] DEAK P, ARADI B, FRAUENHEIM T J J P C C. Band lineup and charge carrier separation in mixed rutile-anatase systems[J]. J. Phys. Chem. C, 2011, 115:3443-3446.

[54] ZHU Q, PENG Y, LIN L, et al. Stable blue TiO_{2-x} nanoparticles for efficient visible light photocatalysts[J]. J. Mater. Chem. A, 2014, 2:4429-4437.

[55] KAMEGAWA T, MATSUURA S, SETO H, et al. A visible-light-harvesting assembly with a sulfocalixarene linker between dyes and a Pt-TiO_2 photocatalyst[J]. Angew. Chem. Int. Ed., 2013, 52:916-919.

第 7 章

硼掺杂诱导两相比例可控二氧化钛的光催化降解特性

7.1 相结与掺杂共同作用的二氧化钛研究进展

阿特拉津是一种被广泛使用的农药，因其高毒性和难以被生物降解的特性备受关注[1-2]。目前，微生物法[3-4]、电化学法[5-7]及光催化法[8-9]被用来处理含阿特拉津的废水。其中，光催化降解法因其设备简单、高效而受到广泛重视[10]。

先前工作已经证明，具有低金红石含量的相结 TiO_2 对污染物具有良好的光催化降解活性，由于相结有利于光生载流子的分离。同时研究显示，为了增强自身的光催化性能，在 TiO_2 中掺杂其他元素是一种简单可行的方法[1]。尤其是掺杂一些非金属元素被认为是一种有效的方法[3]。同样，掺杂一些诸如非金属元素 N、C、S、P 及卤素元素是一种能使 TiO_2 具有可见光活性的潜在方式[4]。在这些掺杂方式中，硼掺杂因其可以增强 TiO_2 接受电子的能力，而被应用于电化学研究之中，硼元素倾向于替代氧原子或者节点间隙的位置。此外，利用密度泛函计算硼掺杂后的电子结构揭示出 B 的 p 轨道和 O 的 2p 轨道发生混合，这使得 TiO_2 的禁带发生收缩[5,10]。

与此同时，调节后的 TiO_2 因其具有优异的光生电子和空穴的分离能力[11]，延长了复合的时间，从而受到广泛的关注[6-7]。

而正如之前的研究报道，在退火的过程中优化调节制备 TiO_2，更多的 Cl 或者 H_2O 在相对较高的 HCl/Ti 比例下可以有效屏蔽正电荷，从而可以降低静电（排斥力）的作用[8]。此外，硼酸作为一种引入 TiO_2 的硼源，可能也具有和 HCl 类似的作用，从而实现调节 TiO_2 各相的比例以及完成 B 掺杂过程。目前制备调节的 TiO_2 的方法需要使用有机试剂，制备过程复杂还需要模板[9,12]。因此，目前需要一种简单的可以优化掺杂的比例的方法，从而实现合理的（A/R）比例，增强 TiO_2 的光催化性能[13]。而阿特拉津作为一种典型的难降解有机污染物，被用来研究合成材料的光催化性能[14-22]。

在本章中，我们使用硼酸和 Ti^{3+} 溶液作为反应试剂，在相转变温度下通过一步煅烧的方法制备 B 掺杂的 TiO_2（A/R）纳米粒子；分析了 B 掺杂在光催化过程中对电子能带结构、光学性质、光生电子-空穴对的分离和转移以及活性氧簇生成的影响，考察了 B 掺杂的 TiO_2 作为光催化剂降解阿特拉津的能力。此外，还探索了光催化降解阿特拉津的途径。

7.2 硼掺杂两相比例可控的二氧化钛的合成与表征

7.2.1 硼掺杂两相比例可控的二氧化钛的制备

实验中使用的试剂纯度均为分析纯,且未进一步处理。

B 掺杂的 TiO_2(A/R)纳米粒子的制备方法如下:先用丙酮、乙醇以及去离子水清洗 0.3 mm 厚的钛箔(0.6 g Ti),然后加入到具有 50 mL 有效体积的聚四氟乙烯内衬不锈钢反应釜中,之后将 20 mL 的水和 20 ml 的浓盐酸混合物缓慢地加入反应釜中,最后,在烘箱中 160 ℃反应 2 h,自然冷却到室温,获得制备好的溶液。然后,将不同剂量的硼酸(0 mg、15 mg、30 mg、60 mg、120 mg 以及 150 mg)分散在 10 mL 制备好的溶液中,然后将得到的混合溶液在 500 ℃的条件下在空气中退火 2 h,其中加热速度为 5 ℃/min,依硼酸不同的加入量得到的产物分别标记为 B_1、B_2、B_3、B_4、B_5 以及 B_6。

7.2.2 硼掺杂两相二氧化钛的物相分析及化学性质

所获得的材料,利用 Rigaku 衍射仪(Rigaku 公司,日本)获得功率 X 射线衍射(XRD)图,使用 Cu K 辐射源($\lambda = 1.541841 \times 10^{-10}$ m),扫描速度为 8°/min,加速电压和电流分别为 40 kV 和 200 mA。实验图 7.1(a)为不同 B 掺杂样品的 XRD 谱图。无掺杂和掺杂后的样品的晶体结构主要是由锐钛矿相($2\theta = 25.4°$)和金红石相($2\theta = 27.5°$)组成。这表明 TiO_2 纳米粒子中包含有两种物相,而各相的比例则可以利用 XRD 的峰强度通过以下公式计算得到[23]:

$$f_A = \frac{1}{1 + \frac{1}{K}\frac{I_R}{I_A}} \qquad (7.1)$$

$$K = 0.79 \quad f_A > 0.2$$

$$K = 0.68 \quad f_A \leqslant 0.2$$

其中,f_A 为粉末中锐钛矿相的比例。I_A 和 I_R 则分别表示的是锐钛矿{101}和金红石(110)X 射线衍射峰的峰强。金红石相在 6 个样品中的重量比分别为51.8 wt%、12.6 wt%、20.9 wt%、17.5 wt%、10.3 wt%以及 7.5 wt%。此外,与非掺杂的 TiO_2 对比,在 XRD 谱图中可以看出,$B_2 \sim B_6$ 中金红石相的比例是下降的,而且,掺杂后的锐钛矿相和金红石相主要的峰位置发生了明显的位移。

我们之前的研究证明了溶液中相对较高的 HCl/Ti 比例可以减少静电排斥的效应,而静电排斥效应则被认为阻碍金红石相的形成[18,24]。此外,硼酸的存在则会促进静电排斥的效应,所以实验中氯化氢和硼酸的协同作用来调节 TiO_2 物相。

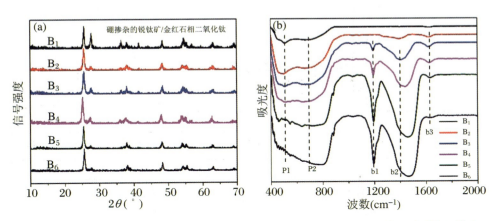

图 7.1 不同硼掺杂量的 B-doped TiO_2(A/R)纳米颗粒的 XRD 图(a)和傅立叶变换红外光谱图(b)

0 mg(B_1)、15 mg(B_2)、30 mg(B_3)、60 mg(B_4)、120 mg(B_5)和 150 mg(B_6)

为了进一步探究两相混合 TiO_2 的物相及 B 的掺杂特性,材料的傅里叶变换红外光谱(FTIR)在 Vertex 70 光谱仪(Bruker 公司,德国)中记录并分析。在红外光谱谱图中,1000 cm^{-1} 以下指带峰为 Ti—O—Ti 网络的特征峰,可以证明 B 掺杂 TiO_2 的存在,三个强度信号分别为 1190 cm^{-1}、1400 cm^{-1} 以及 1620 cm^{-1}。而且可以很明显观察到 694 cm^{-1} 和 500 cm^{-1} 的两个峰演变成一个 690 cm^{-1} 的单峰,如图 7.1(b)中 P1、P2 所示。为了解析这两个峰的起源,我们探究了两相混合 TiO$_2$ 的红外光谱。锐钛矿型 TiO$_2$ 在 694 cm^{-1} 处有一个单峰,而金红石型 TiO$_2$ 则具有在 656 cm^{-1} 处和 528 cm^{-1} 处的两个峰,所以图 7.1(b)中 525 cm^{-1} 处的峰是来自于金红石相的二氧化钛,而 664 cm^{-1} 处的主峰则是锐钛矿和金红石相结构结合导致[25]的。

除了 1000 cm^{-1} 以下的峰,图中也可以观察到 1190 cm^{-1}、1400 cm^{-1} 和 1620 cm^{-1} 的峰。1620 cm^{-1} 的振动峰是表面吸附的水以及羟基导致(表示为 b3)的,而 1400 cm^{-1} 处的峰则是由于硼中三价的 B 的出现(以 B 3p 的形式,表示为 b2),趋向于与环境中的氧原子相互作用,表现出与正常的 Ti—O—B 相似的化学环境,但是在 1190 cm^{-1} 处的峰应该对应的是 B—O 键(表示为 b1)的伸缩振动[26]。这些结果与 XRD 的结果有很好的一致性。简单地说,上述讨论的红外结构表明掺杂在 TiO$_2$ 中的硼包含三价硼的结构,然而,并不能排除氧化硼是否存在。

除了红外光谱分析,实验中化学成分表征则通过 X 射线光电子谱分析 (XPS,Ulvac-Phi 公司,日本)来探究纳入 TiO$_2$ 物相中掺杂剂的化学价态。图

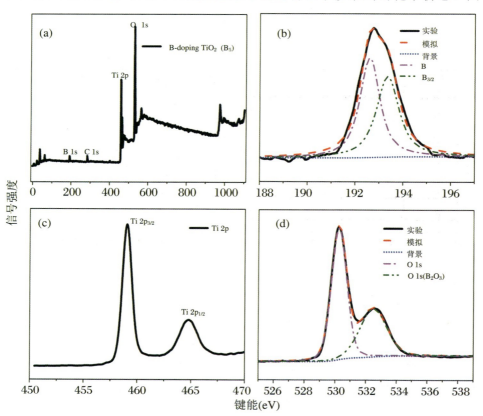

图 7.2　B-doped TiO$_2$(A/R)(B$_5$)样品的 XPS 光谱:(a) 全谱;(b)B 1s;(c) Ti 2p;(d) O 1s

7.2(a)为样品 B_5 的 B 1s、O 1s 以及 Ti 2p 的核水平能量。硼的 1s XPS 谱图表明 TiO_2 表面掺杂上了硼元素。如图 7.2(b)所示,硼的 1s XPS 谱图中在 191.0~195.0 eV 区间出现了典型不对称宽峰,这也许是 192.6 eV 和 193.3 eV 两重峰的重叠,192.6 eV 处的峰可能归因于硼的间隙,而 193.3 eV 则可能源于 $BO_{3/2}$ 聚合物和表面 $BO_{3/2}$ 物质的信号[16,26]。另外,在红外谱图中,也可以对 B 做类似的分析。此外,相对于 $BO_{3/2}$,光谱分析表明间隙硼在总硼含量中占据更大比例,这与红外光谱分析一致。

硼掺杂 TiO_2 中钛的 $2p_{3/2}$ XPS 在 459.0 eV 出现高峰,在 464.7 eV 为 $2p_{1/2}$ 峰,此外,在图 7.2(d)中,O 1s(B_2O_3)有一个 $BO_{3/2}$ 物质的峰。

7.2.3

硼掺杂两相二氧化钛纳米材料的微观结构分析

接着,通过透射电子显微镜(TEM,JEOL 公司,日本)对样品的形态和结构进行表征。图 7.3 为加入不同浓度硼酸制备得到的硼掺杂 TiO_2 的低倍透射电镜图。对比的纳米复合物(B_1)具有很多的纳米棒。当硼的加入量逐渐增大,获得的样品为很多的纳米粒子,只会出现很少的纳米棒。所有的样品显示出相似的形貌,如不同尺寸的 TiO_2 纳米粒子和纳米棒的聚集体。

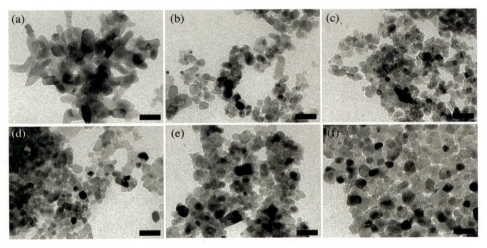

图 7.3　B-doped TiO_2(A/R)样品典型的透射电镜图:(a) B_1;(b) B_2;(c) B_3;(d) B_4;(e) B_5;(f) B_6

标尺:60 nm

为了能提供直接的证据去证明反应产物中含有锐钛矿和金红石两种物相，实验对制备的硼掺杂 TiO_2 进行了高倍透射电子显微镜（HRTEM，JEOL 公司，日本，以 200 kV 的加速电压进行高分辨率透射电子显微镜图像）测试。图 7.4 为 B_5 的高倍透射电子显微镜图像，可以明显地看到，晶格间距为 0.32 nm，对应的为金红石相的（110）面。而相邻的纳米粒子的晶格间距为 0.35 nm，对应的是锐钛矿相的{101}面。因此，高倍透射电子显微镜进一步证明了 TiO_2 纳米粒子中共同存在锐钛矿相和金红石相，而且也可以清楚地看到两相之间的结合。

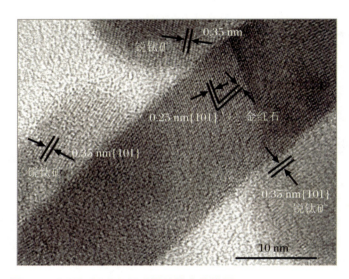

图 7.4　B-TiO_2（A/R）（B_5）样品的高分辨图

7.2.4
硼掺杂两相二氧化钛纳米材料的光物理性质及光化学性质

半导体材料的光吸收能力与材料自身的电子能带结构密切相关。样品的漫反射光谱（DRS）测量通过 UV-vis 分光光度计（Shimadzu 公司，日本）获得。图 7.5 是硼掺杂对光吸收特性的影响。图 7.5 的插图是 350 nm 到 545 nm 的放大图。所有硼掺杂 TiO_2 边缘的位置发生了很大的蓝移，表明硼掺杂 TiO_2 的禁带变宽，这与 BO_x 的禁带大于 TiO_2 有关。这是由于掺杂的硼元素位于 TiO_2 结构的间隙，从而形成了一个稳定的化学环境，比如 Ti—O—B[27-28]。此外，紫外-可见边缘发生了明显的蓝移现象可能是由于掺杂后金红石相的减少导致[29]的。

因此，硼掺杂的 TiO_2 的吸光率是由金红石和硼元素共同导致的。这也许可以解释为什么 B_2 相比于 B_3 和 B_4 有着更大的蓝移。此外，相对于未掺杂的 TiO_2，掺杂 TiO_2 的光吸收能力会加强[30]。这增强了材料在紫外-可见区域的光催化活性。

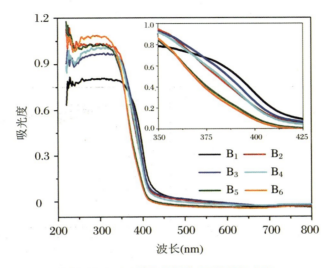

图 7.5　所有 TiO_2(A/R)样品的紫外-可见吸收光谱

7.3 硼掺杂两相二氧化钛纳米材料的光催化阿特拉津降解性能及降解路径分析

为了评估硼掺杂 TiO_2 的光活性，我们进行了光催化降解阿特拉津实验（图 7.6）。实验探究了不同产物的阿特拉津光催化降解能力。实验中，将 20 mg 未掺杂及掺杂的 TiO_2 分别分散在 40 mL 的阿特拉津（10 mg/L）中，并在黑暗中搅拌 30 min 以使催化剂和反应溶液达到吸附解吸附平衡。之后，在光催化实验中，使用具有 300 nm 的滤光片的 350 W（15 A）氙灯（CHFXM-350W，Beijing Trusttech 公司，中国）对所有污染物溶液进行辐照。光催化降解实验使用 50 mL 的烧杯，在冰水浴的条件下进行。实验取样为在反应液中取 1 mL 的溶液过滤得到。空白对照（不加入光催化剂）组表面阿特拉津在紫外-可见辐照下浓度保持稳定。而 B 掺杂 TiO_2 存在下，阿特拉津的浓度发生了快速下降，其中，B_5 在光照 180 min 时有着最高的降解效率，为 94.7%。而非掺杂的 TiO_2 样品对阿特拉津的降解率仅仅为 50.1%。对于所有样品，当金红石相的比重逐渐下降到 11.6wt%时，光催化降解阿特拉津的效率会逐渐升高。这个结果表明在两相混合的 TiO_2 中，金红石、锐钛矿的比例在控制光生电子-空穴对的复合中起到了至关重要的作用，从而影响它的光催化活性。

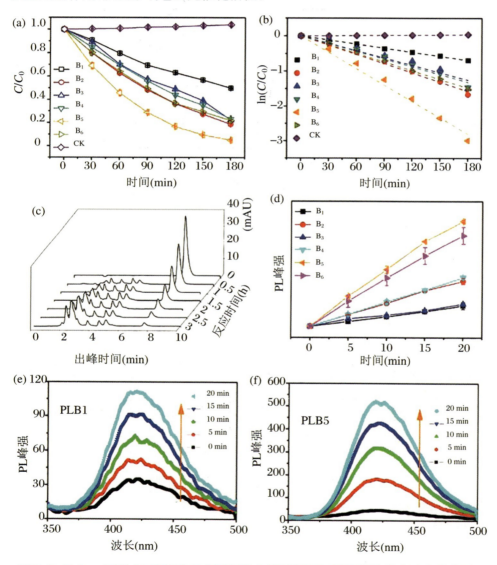

图 7.6　B-doped TiO_2(A/R)所有样品在紫外-可见光照射下的表征结果：(a) 光催化降解对照图；(b) 相应的伪一级动力学常数拟合；(c) 光催化降解阿特拉津中 HPLC 色谱谱；(d~f) 随反应时间变化在 426 nm 处的荧光强度图

实验中利用 HPLC(HPLC-1100, Agilent 公司,美国)测试得到阿特拉津的浓度,移动相则是水和甲醇的混合液(体积比为 40∶60),流速为 0.8 mL/min。图 7.6(c)为在 B_5 条件下阿特拉津随时间的降解过程,图中可以看到中间产物在 3~6 min 发生聚集,然后因为降解速率大于聚集速度而发生下降。图 7.6(b)为阿特拉津随时间的降解过程,由于中间体的降解过程使其符合假一级动力学[31]。因此,阿特拉津的表观降解速率常数(k)可以通过 $\ln(C/C_0)$ 和时间的比值拟合得到,其中,C 为阿特拉津的浓度,不同催化剂之间的阿特拉津降解速率常数比较表明实现最快降解速率的为两相混合的 TiO_2。此外,k_{B_5} 相对于其他催化剂在绝对数值上最大,这表明在相结 TiO_2 中优化硼掺杂量来提高光催化降解效果是一个可行的方法。

在光催化过程中,TiO_2 吸收光产生电子-空穴对,它可以和吸附在催化剂表面的氧气和水发生反应,生成强氧化性的羟基自由基。实验中,羟基自由基是降解有机污染物的主要物质。因此,基于羟基自由基可以和对苯二甲酸反应生成二羟基对苯二甲酸,得到的产物在 426 nm 会发射出特有的荧光。我们在实验中利用荧光光谱仪(LS 55,Perkin Elmer 公司,美国),在室温条件下,用 310 nm 激发波长来检测羟基自由基的浓度。从图 7.6(d)中可以很明显地看出在新形成的价带上产生的空穴会快速地转移到 TiO_2 包面吸附的羟基和水上,从而形成羟基自由基。同时,从图 7.6(e,f)荧光强度和辐照时间之间很好的线性关系表明均一硼掺杂 TiO_2 有良好的稳定性。此外,除了主要是空穴攻击水和羟基的方式路线,在随后的反应中,也有另外一种可能产生羟基自由基的方式。

实验利用液相/质谱分析(LC/MS,6460,Agilent 公司,美国)进一步检测反应中间产物的降解,以探究硼掺杂 TiO_2 光催化降解阿特拉津的路径。总结实验的结果和参考文献报道[32-34],我们提出了以下几种硼掺杂 TiO_2 降解阿特拉津的途径,如图 7.7 所示。

2-羟基阿特拉津,2-氯-4-乙胺基-6-(1-甲基-1-乙醇)胺基-1,3,5-三嗪以及 desethyldesisopropyl atrazine 在测试中并未检测到,这可能是由于它们自身浓度较低或者不稳定。其余在 LC-MS 检测后分析所得如表 7.1 所示。结果与文献报道符合[35-36]。

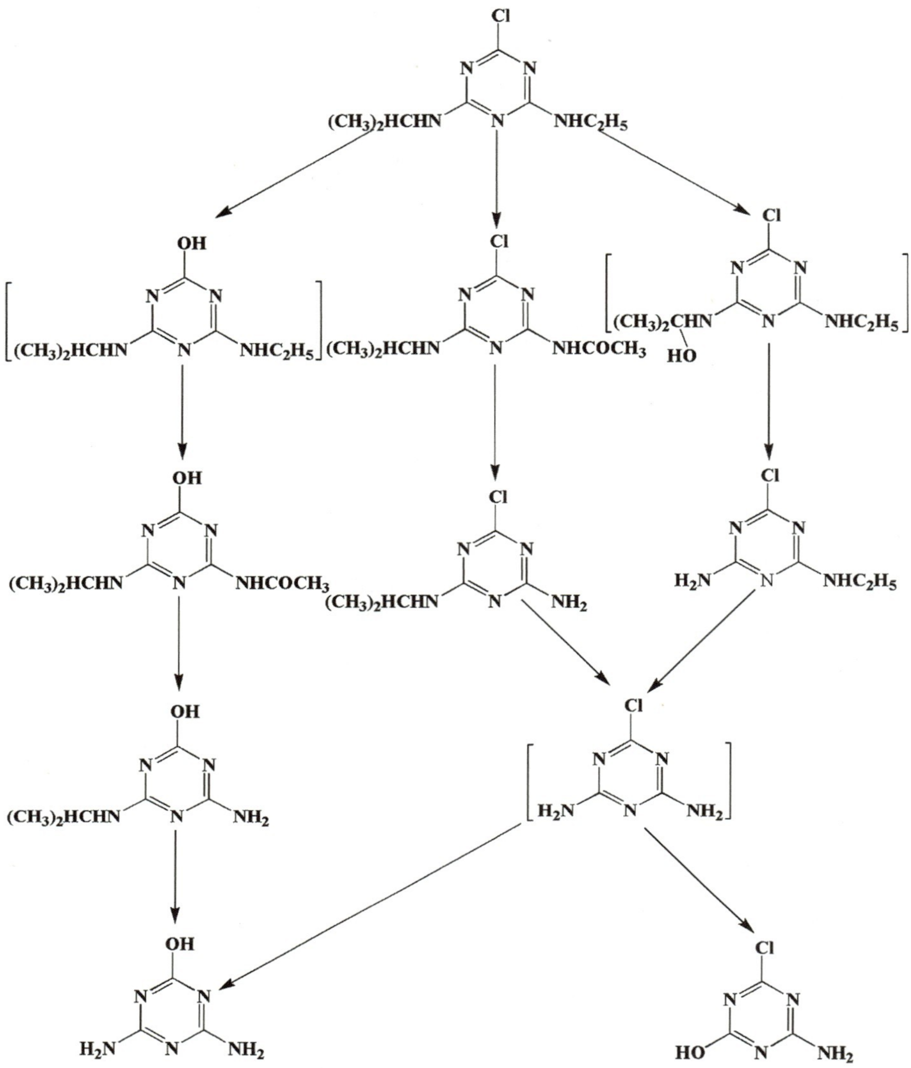

图 7.7　在 UV-vis 照射下，污染物在 B-doped TiO_2（A/R）催化下的可能降解路径

表 7.1　B-doped TiO_2(A/R)光催化降解阿特拉津过程中的中间产物及其在液相中的保留时间

序号	简称	化合物名称	检测时间（min）
I	ATRAZINE	Atrazine	26.88
II	OHA	2-Hydroxyatrazine	undetected
III	AOHE	2-Chloro-4-acetamindo-6-isopropylamino-1,3,5-triazine	13.53
IV	AOHI1	2-Chloro-4-ethylamino-6-(1-methyl-1-ethanol)amino-1,3,5-triazine	undetected
V	OHOE	2-Hydroxy-4-acetamido-6-isopropylamino-1,3,5-triazine	5.15
VI	DEA	Desethylatriazine	8.70
VII	DIA	Deisopropylatriazine	6.09
VIII	OHDEA	2-Hydroxydesethyl atrazine	4.36
IX	DAA	Desethyldesisopropyl atrazine	undetected
X	AME	2-Hydroxy-4,6-diamino-1,3,5-triazine	3.70
XI	ClOHNH$_2$	2-Chloro-4amino-6-hydroxy-1,3,5-triazine	4.61

7.4 硼掺杂两相二氧化钛纳米材料的光催化机制分析

3 个原因可能对优化的 B 掺杂的 TiO_2(A/R)-B$_5$ 的光催化性能具有提高作用。

首先，在锐钛矿和金红石颗粒之间形成的相结能够极大地促进光催化过程中的光催化活性[37-38]。相结的存在有利于光激发产生的载流子（电子-空穴对）的分离，进一步提高光催化活性[37]。众所周知，包括 TiO_2 在内的光催化剂（或其他半导体)通过吸收一个光子被激发,光子的能量大于半导体的能带间隙,激发后产生电子-空穴对。如图7.8所示,这种形成的能带结构能够促进光生电子在界面处从金红石转移到锐钛矿,然后空穴从锐钛矿转移到金红石[39]。

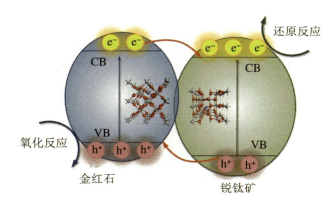

图7.8 硼掺杂锐钛矿与金红石界面可能的导带与价带排列

其次,载流子的捕获位点也随着 B 掺杂量的增加而增加。光生载流子寿命的延长有利于光激发产生的电子-空穴的分离。除此之外,在光催化过程中,B 表面促进电子-空穴对利用吸收的水分子产生·OH 自由基。与未掺杂的 TiO_2 相比,B 掺杂 TiO_2(A/R)之后价带位置向正向偏移,从而有利于·OH 的产生,大幅度提高 TiO_2 的光催化氧化能力[40]。表面羟基官能团的形成有利于提高光催化活性,因为它们能够调节氧化的电子转移。在光照射之后,导带产生电子,这时晶体间隙的 B 作为浅阱能够捕获电子,进而延长光生载流子的寿命。这个过程对提高 B 掺杂的 TiO_2 的紫外-可见光的光催化活性是必不可少的。

最后,e^- 和 h^+ 能够扩散到光催化剂的表面,然后继续和羟基官能团,吸附水和氧气反应,形成高活性的·OH。同时,表面的 B 能够在 B 掺杂的 TiO_2 表面引入布朗斯特和路易斯酸中心[26],这也有利于提高 TiO_2 的光催化活性。

上述的这些原因能够促进 B 掺杂的 TiO_2 成为一种有效的光催化剂。因此,硼掺杂不仅能够调节混合相中金红石的含量,掺杂自身也能够提高催化剂的光催化活性。

以前一章的研究为基础,本章中我们在 Ti 源的溶液中通过掺入硼酸作为前驱,在相转化温度下通过一步煅烧的方法合成 B 掺杂的金红石/锐钛矿两相 TiO_2。理论证明,掺杂和金红石/锐钛矿相结都对光生载流子的分离起着重要作用,它们可以分别作为电子的捕获位点和驱动力。实验结果证明,该设计合成的 B 掺杂的金红石/锐钛矿相结 TiO_2(金红石含量为 10.3 wt%,以 120 mg 硼酸为掺杂原料)在紫外和可见光下降解阿特拉津时表现最高的降解效率,为 94.7%,其光催化降解速率是未掺杂的 TiO_2(降解效率 50.1%)的 4 倍。该工作为掺杂其他元素的金红石/锐钛矿两相 TiO₂ 提供了可行的方法，且说明该催化剂在紫外-可见光下可以降解多种难降解污染物，在水处理领域中有较好的应用前景。

参考文献

[1] JAISWAL R, PATEL N, DASHORA A, et al. Efficient Co-B-codoped TiO₂ photocatalyst for degradation of organic water pollutant under visible light [J]. Appl. Catal. B., 2016, 183:242-253.

[2] ESPARZA-NARANJO S B, DA SILVA G F, DUQUE-CASTANO D C, et al. Potential for the biodegradation of atrazine using leaf litter fungi from a subtropical protection area [J]. Curr. Microbiol., 2021, 78:358-368.

[3] PATEL N, JAISWAL R, WARANG T, et al. Efficient photocatalytic degradation of organic water pollutants using V-N-codoped TiO₂ thin films [J]. Appl. Catal. B., 2014, 150:74-81.

[4] SU Y, HAN S, ZHANG X, et al. Preparation and visible-light-driven photoelectrocatalytic properties of boron-doped TiO₂ nanotubes [J]. Mater. Chem. Phys., 2008, 110: 239-246.

[5] ZHAO W, MA W, CHEN C, et al. Efficient degradation of toxic organic pollutants with $Ni_2O_3/TiO_{2-x}B_x$ under visible irradiation [J]. J. Am. Chem. Soc., 2004, 126: 4782-4783.

[6] LIU J, YU X, LIU Q, et al. Surface-phase junctions of branched TiO₂ nanorod arrays for efficient photoelectrochemical water splitting [J]. Appl. Catal. B., 2014, 158:296-300.

[7] WANG W K, CHEN J J, ZHANG X, et al. Self-induced synthesis of phase-junction TiO₂ with a tailored rutile to anatase ratio below phase transition temperature [J]. Sci. Rep., 2016, 6: 20491.

[8] LI G, GRAY K A. Preparation of mixed-phase titanium dioxide nanocomposites via solvothermal processing [J]. Chem. Mater., 2007, 19: 1143-1146.

[9] ZUKALOV M, ZUKAL A, KAVAN L, et al. Organized mesoporous TiO₂ films exhibiting greatly enhanced performance in dye-Sensitized solar cells [J]. Nano Lett., 2005, 5:1789-1792.

[10] HABIBI R, GILANI N, PASIKHANI J V, et al. Improved photoelectrocatalytic activity of anodic TiO_2 nanotubes by boron in situ doping coupled with geometrical optimization: application of a potent photoanode in the purification of dye wastewater [J]. J. Solid State Electr., 2021, 25: 545-560.

[11] YANG X, SUN H, LI G, et al. Fouling of TiO_2 induced by natural organic matters during photocatalytic water treatment: mechanisms and regeneration strategy [J]. Appl. Catal. B., 2021, 294:120252.

[12] CHEN S-L, WANG A-J, HU C-T, et al. Enhanced photocatalytic performance of nanocrystalline TiO_2 membrane by both slow photons and stop-band reflection of photonic crystals [J]. AIChE J., 2012, 58:568-572.

[13] WANG Y, LIU X, GUO L, et al. Metal organic framework-derived C-doped ZnO/TiO_2 nanocomposite catalysts for enhanced photodegradation of Rhodamine B [J]. J. Colloid Interface Sci., 2021, 599:566-576.

[14] PAPPAS E A, HUANG C. Predicting atrazine levels in water utility intake water for MCL compliance [J]. Environ. Sci. Technol., 2008, 42: 7064-7068.

[15] WACKETT L, SADOWSKY M, MARTINEZ B, et al. Biodegradation of atrazine and related s-triazine compounds: from enzymes to field studies [J]. Appl. Microbiol. Biot., 2002, 58:39-45.

[16] SILVA E, FIALHO A M, S -CORREIA I, et al. Combined bioaugmentation and biostimulation to cleanup soil contaminated with high concentrations of atrazine [J]. Environ. Sci. Technol., 2004, 38:632-637.

[17] CHEN X, HU X, AN L, et al. Electrocatalytic dechlorination of atrazine using binuclear iron phthalocyanine as electrocatalysts[J]. Electrocatalysis, 2014, 5:68-74.

[18] VENTURA A, JACQUET G, BERMOND A, et al. Electrochemical generation of the Fenton's reagent: application to atrazine degradation [J]. Water Res., 2002, 36:3517-3522.

[19] MALPASS G R P, MIWA D W, MACHADO S A S, et al. Oxidation of the pesticide atrazine at DSA© electrodes [J]. J. Hazard. Mater., 2006, 137: 565-572.

[20] CHEN H, YANG S, YU K, et al. Effective photocatalytic degradation of atrazine over titania-coated carbon nanotubes (CNTs) coupled with micro-

wave energy [J]. J. Phys. Chem. A., 2011, 115: 3034-3041.

[21] ZHANQI G, SHAOGUI Y, NA T, et al. Microwave assisted rapid and complete degradation of atrazine using TiO_2 nanotube photocatalyst suspensions[J]. J. Hazard. Mater., 2007, 145: 424-430.

[22] MAHLALELA L C, CASADO C, MARUG N J, et al. Coupling biological and photocatalytic treatment of atrazine and tebuthiuron in aqueous solution [J]. J. Water Process. Eng., 2021, 40: 101918.

[23] LI G, RICHTER C P, MILOT R L, et al. Synergistic effect between anatase and rutile TiO_2 nanoparticles in dye-sensitized solar cells [J]. Dalton Trans., 2009, 45: 10078-10085.

[24] ZHENG W, LIU X, YAN Z, et al. Ionic liquid-assisted synthesis of large-scale TiO_2 nanoparticles with controllable phase by hydrolysis of $TiCl_4$ [J]. ACS nano, 2009, 3: 115-122.

[25] LIU G, YAN X, CHEN Z, et al. Synthesis of rutile-anatase core-shell structured TiO_2 for photocatalysis [J]. J. Mater. Chem., 2009, 19: 6590-6596.

[26] FENG N, ZHENG A, WANG Q, et al. Boron environments in B-doped and (B, N)-codoped TiO_2 photocatalysts: a combined solid-State NMR and theoretical calculation study [J]. J. Phys. Chem. C., 2011, 115: 2709-2719.

[27] CHEN D, YANG D, WANG Q, et al. Effects of boron doping on photocatalytic activity and microstructure of titanium dioxide nanoparticles [J]. Ind. Eng. Chem. Res., 2006, 45: 4110-4116.

[28] GUO M, DAI Y, HUANG B. Tailoring the band gap of GaN codoped by VO for enhanced solar energy conversion from first-principles calculations [J]. Comput. Mater. Sci., 2012, 54: 101-104.

[29] JIN H, DAI Y, WEI W, et al. Density functional characterization of B doping at rutile TiO_2 {110} surface [J]. J. Phy. D., 2008, 41: 195411.

[30] LU N, ZHAO H, LI J, et al. Characterization of boron-doped TiO_2 nanotube arrays prepared by electrochemical method and its visible light activity[J]. Sep. Purif. Technol., 2008, 62: 668-673.

[31] AN T C, ZHU X H, XIONG Y. Feasibility study of photoelectrochemical degradation of methylene blue with three-dimensional electrode-photocatalytic reactor [J]. Chemosphere, 2002, 46: 897-903.

[32] H QUET V, GONZALEZ C, LE CLOIREC P. Photochemical processes for

atrazine degradation: methodological approach [J]. Water Res., 2001, 35: 4253-4260.

[33] LACKHOFF M, NIESSNER R. Photocatalytic atrazine degradation by synthetic minerals, atmospheric aerosols, and soil particles [J]. Environ. Sci. Technol., 2002, 36: 5342-5347.

[34] BALCI B, OTURAN N, CHERRIER R, et al. Degradation of atrazine in aqueous medium by electrocatalytically generated hydroxyl radicals. A kinetic and mechanistic study [J]. Water Res., 2009, 43: 1924-1934.

[35] HEQUET V, GONZALEZ C, LE CLOIREC P. Photochemical processes for atrazine degradation: methodological approach [J]. Water Res, 2001, 35: 4253-4260.

[36] LACKHOFF M, NIESSNER R. Photocatalytic atrazine degradation by synthetic minerals, atmospheric aerosols, and soil particles [J]. Environ. Sci. Technol., 2002, 36:5342-5347.

[37] ZHANG J, XU Q, FENG Z, et al. Importance of the relationship between surface phases and photocatalytic activity of TiO_2 [J]. Angew. Chem. Int. Ed., 2008, 47: 1766-1769.

[38] KIM W, TACHIKAWA T, MOON G-H, et al. Molecular-level understanding of the photocatalytic activity difference between anatase and rutile nanoparticles [J]. Angew. Chem. Int. Ed., 2014, 53: 14036-14041.

[39] SCANLON D O, DUNNILL C W, BUCKERIDGE J, et al. Band alignment of rutile and anatase TiO_2 [J]. Nat. Mater., 2013, 12: 798-801.

[40] WU T T, XIE Y P, YIN L C, et al. Switching photocatalytic H_2 and O_2 generation preferences of rutile TiO_2 microspheres with dominant reactive facets by boron doping [J]. J. Phys. Chem. C., 2015, 119: 84-89.

第 8 章

氧化锌/硫化镉/二氧化钛异质结光催化剂的构筑及其光催化降解特性

8.1 二氧化钛与金属氧化物/硫化物异质结的研究进展

根据文献综述,原则上,TiO_2可以当作完美的催化剂[1-6],但是由于其固有的原子结构和电子结构,两个主要障碍在很大程度上限制了其在太阳辐照下的环境应用:宽带隙(3.0～3.2 eV)和低电荷迁移率[3,7-12]。为了克服这两个障碍,TiO_2与窄禁带半导体的结合对于低成本实际应用具有相当大的价值[4-5,11,13]。由于CdS具有窄的禁带隙(2.2～2.4 eV)、高的吸收系数和合适的导带,它被优先用于敏化TiO_2的可见光光活性[8,14-20]。为了进一步提高光子转换效率和表观催化性能,基于高度有序、垂直取向的纳米管阵列的CdS/TiO_2复合材料由于能增强电荷转移而引起了越来越多的关注[11,15,21-22]。但是这种复合材料在实际污染物降解过程中,特别是在太阳光辐射下的水溶液中,其固有的光腐蚀行为很大程度上限制了它的应用[15,21,23]。这一缺点可以用牺牲试剂清除硫化镉中的腐蚀孔来克服[14,24-25],例如Na_2S/Na_2SO_3[14]、S^{2-}/SO_3^{2-}、I^-/I_3^-、Co^{2+}/Co^{3+}[26]、Fe^{2+}/Fe^{3+}[27]、乙醇[14]、乳酸[18]、三乙醇胺[28]、抗坏血酸[29]和复合氧化还原电对[30]。实际上,牺牲试剂作为电子供体,应该具有足够低的电位,并且可以很容易地被低电容($E^0=1.38$ V)的CdS空穴氧化,然后对氧化后的CdS进行再生和回收。但是,这些牺牲试剂由于成本高、消耗大、副反应和二次污染等原因,不适合实际应用[14,31-32]。因此,具有较大环境优势的固体牺牲试剂是非常有前景的。

近年来,为了解决上述问题,提出了基于直接Z型异质结材料[19,33-36]。在这种复合物中,具有低氧化活性的空穴在复合界面与具有低还原活性的电子快速复合,而具有高氧化活性的空穴和具有高还原活性的电子则被有效地利用。因此,在这种情况下不需要外加牺牲试剂[37-38]。CdS/ZnO复合物可以有效地构建这样一种直接Z型结构,并具有较高的催化反应效率[39-41]。在这种复合体系中,CdS通常沉积在ZnO上,太阳光沿着CdS照射到ZnO。然而,由于CdS对紫外和可见光子的吸收占主导地位,这种结构可能限制ZnO的活化,从而降低Z型异质结的整体性能。因此,如果ZnO沉积在CdS上形成ZnO/CdS复合物,太阳光会沿ZnO照射到CdS,此时由于ZnO和CdS分别选择性地吸附紫外光、可见光,ZnO的活化和复合结构的整体性能可以显著提高。此外,充分活化的ZnO也可以为克服CdS光腐蚀问题提供一个不用外加牺牲试剂的解决方案。

在本章中，我们设计并制备了一种用于光催化降解污染物的ZnO/CdS/TiO_2异质结材料。在这种复合材料中，内层的TiO_2纳米管具有较大的比表面积、较强的捕光能力、便捷的电荷转移方式和较高的电荷收集效率，从而提高了光催化活性，而外层的ZnO作为中间层CdS的固体牺牲试剂，提高了抗光腐蚀的能力。该材料制备后，从形貌、催化能力和稳定性等方面对其进行了表征，并进一步用于降解两种典型的有机污染物（阿特拉津和罗丹明B），以及在模拟太阳光照射下处理纺织印染实际废水。本工作为研制新型抗光腐蚀的异质结光催化剂，用于可见光催化降解难降解有机污染物提供了一种新的方法。

8.2 氧化锌/硫化镉/二氧化钛异质结的合成与表征

8.2.1 氧化锌/硫化镉/二氧化钛异质结的制备

本章中所使用的试剂均为分析纯，且未作任何深度纯化。根据图8.1的制备流程图，首先，0.30 mm厚的钛箔需要依次用丙酮、乙醇和去离子水清洗。接着，将0.30 mm的钛箔抛光并浸入抛光液（HF：HNO_3：H_2O=1：1：2，体积比）中以去除氧化物层和污渍，并在化学抛光后分别用丙酮、乙醇和水清洗。通过阳极氧化法制备TiO_2纳米管阵列（TiO_2-NTs，2 cm×3 cm），其中两电极结构由钛箔作为阳极和另一个Ti箔作为阴极组成。随后，在0.09 mol/L的NH_4F、去离子水和乙二醇的搅拌溶液中进行阳极氧化。支持电解液中的含水量和阳极电压分别在1%～10%（体积比）和30～90 V范围内。电化学处理从开路电位到终端电位,然后在 15~25 ℃条件下保持电位常数 4~5 天。获得的 TiO$_2$-NTs 前驱体依次用乙醇和去离子水清洗,在 60 ℃的烘箱中烘干。最后样品在马弗炉中煅烧,在 450 ℃保持 2 h,自然冷却后,获得 TiO$_2$-NT 样品。

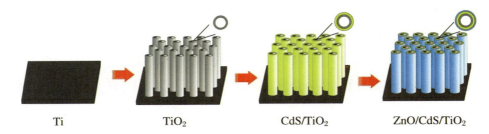

图 8.1　ZnO/CdS/TiO$_2$ 异质结的制备流程图

采用氯化镉、三乙醇胺、氨水和硫脲等化学溶液在 TiO$_2$-NTs 表面沉积 CdS-NPs(CdS 纳米颗粒)。将 8 mL 的氯化镉溶液(0.1 mol/L)、4 mL 三乙醇胺溶液(50%)、4 mL 氨水(15 mol/L)和 8 mL 硫脲(1 mol/L)依次加入 80 mL 烧杯中搅拌,然后加入 54 mL 70 ℃的热水。将 TiO$_2$ 纳米管阵列垂直浸入 70 ℃的水浴中,沉积 5 min 后得到均匀明亮的黄色薄膜。然后,将复合物在 60 ℃的真空烘箱中干燥 0.5 h。烘干后,在 500 ℃的 Ar 气中煅烧 1 h,获得 CdS-NPs/TiO$_2$-NTs 样品。

为了进一步制备 ZnO-NPs/CdS-NPs/TiO$_2$-NTs 三元异质结,首先,将含有 1.46 g 无水醋酸锌的 50 mL 乙醇放入冷凝回流装置中,加热至 80 ℃,保持 4 h,得到 ZnO 前驱体。随后,ZnO 前驱体立即用乙醇稀释至 70 mL,在冰水浴中迅速冷却。接着,在稀释冷却后的 ZnO 前驱体中加入 0.29 g 的氢氧化锂,然后将混合液在 0 ℃超声水解 1 h。为了将 ZnO-NPs 沉积固定在 CdS-NPs/TiO$_2$-NTs 上,将未煅烧过的 CdS-NPs/TiO$_2$-NTs 在冷却的 ZnO 前驱体中浸泡足够长的时间,然后在 60 ℃的真空烘箱中干燥 3 h,最后在 500 ℃的 Ar 气中煅烧 1 h,获得 ZnO-NPs/CdS-NPs/TiO$_2$-NTs 样品。

8.2.2

氧化锌/硫化镉/二氧化钛异质结的微观结构分析

完成材料的合成后,利用场发射扫描电子显微镜(FESEM)对其形貌和结构进行了成像分析。图 8.2(a,b)显示制备的高度有序、直立取向的 TiO$_2$-NTs 具有较大的管径、较宽的管间间隙和足够的管长,从而在几何上有利于后续材料的负载。在 CdS 沉积后，纳米管状整体保持基本不变，纳米管的内外表面都变得粗糙、壁更厚、直径更小（图 8.2(c、d)）。这些结果表明，TiO_2-NTs 的一维有序结构没有任何损伤，且 CdS 均匀沉积在 TiO_2-NTs 的内外壁表面。在此基础上，即使在 ZnO 沉积后，仍然可以从它们的横截面视图中识别出一维管状结构，内部和外部表面都变得更加粗糙（图 8.2(e、f)）。与先前大多数研究不同，沉积的敏化材料由于该 TiO_2-NTs 的独特特性而表现出两个显著的特征：① 敏化材料主要沿 TiO_2 纳米管分布，而不是停留在管口；② 不仅分布在纳米管外侧，在管内也存在沉积。因此，在不堵塞管道入口的情况下，CdS 和 ZnO 沿着整个 TiO_2 纳米管支架均匀地沉积在其外部和内表面，这对于催化剂的有效活性位点暴露至关重要。

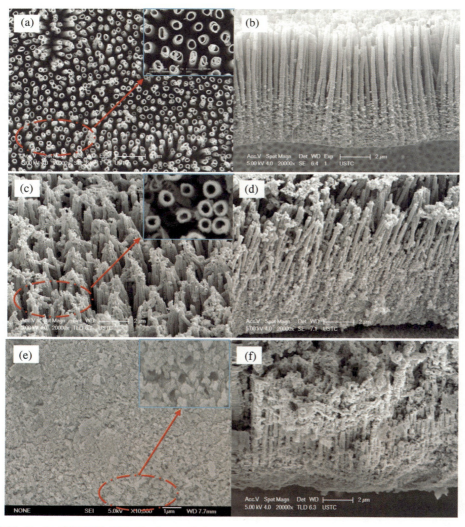

图 8.2 合成样品的 SEM 图：(a、b) TiO_2-NTs；(c、d) CdS-NPs/TiO_2-NTs；(e、f) ZnO-NPs/CdS-NPs/TiO_2-NTs

为了进一步更准确直观地展现该三种材料的分布情况,带球差的高分辨透射电子显微镜(STEM)被用于该异质结的元素定性、元素分布测定及物相的高分辨晶格条纹测量。能量色散 X 射线光谱(图 8.3(g))表明该异质结主要由 Zn、Cd、Ti、O、S 组成,且元素分布图(图 8.3(b~f))进一步说明了 ZnO 和 CdS 在整个 TiO_2 纳米管上按负载顺序分布均匀(图 8.3)。这些结果共同表明成功构建了 $ZnO/CdS/TiO_2$ 异质结。此外,HRTEM 的结果进一步表明,复合材料由 ZnO、CdS 和 TiO_2 组成,它们的特征晶格条纹如图 8.3(g)插图所示。

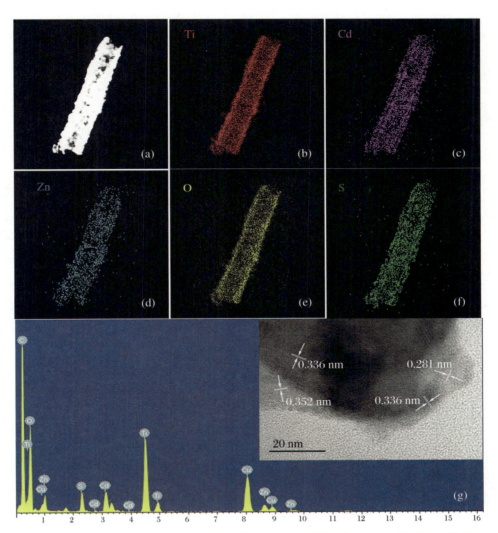

图 8.3 (a) $ZnO/CdS/TiO_2$ NT 的球差电镜高角环形暗场像图像;(b~f) 相应元素的分布图;(g) 对应 EDX 光谱(插图为 HRTEM 图像)

8.2.3

氧化锌/硫化镉/二氧化钛异质结的物相分析及化学性质

通过 X 射线光电子能谱(XPS)对该异质结的化学组成进行测定及分析,图 8.4 清晰地显示了该异质结表面有 Zn、Cd 和 Ti 元素的出峰位置,并给出了具体的结合能参数,与透射电镜的元素定性结果相符合。同时,X 射线衍射(XRD)被用于对该异质结的晶体结构进行分析。图 8.5(a)分别显示了 TiO$_2$-NTs,CdS/TiO$_2$,ZnO/CdS/TiO$_2$ 的物相,结果表明,随着 CdS 和 ZnO 的负载成功,XRD 分别出现了对应 CdS 和 ZnO 的 X 射线衍射峰,从一方面也证明了该三元异质结的成功构筑。

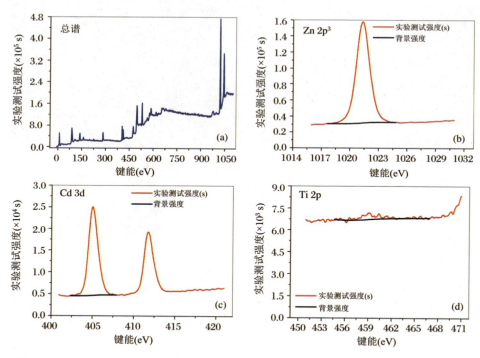

图 8.4 (a) ZnO/CdS/TiO$_2$ 的 XPS 全谱;(b~d) Zn、Cd 和 Ti 元素的高分辨率 XPS 谱图

8.2.4

氧化锌/硫化镉/二氧化钛异质结的光物理性质及光化学性质

为了证明最初设计结构的光学特性,对其在紫外/可见分光光度计上进行了漫反射光谱(DRS)测定。在 ZnO/CdS/TiO$_2$ 复合材料的 DRS 谱中(图 8.5(b)),清晰地观察到两个明显的吸收峰,一个是表层 ZnO 的紫外光吸收,另一个是中间层 CdS 的可见光吸收。此外,ZnO 在 CdS/TiO$_2$ 复合材料上的沉积可以显著改善紫外和可见范围内的光学吸收(图 8.5(b))。事实上,ZnO/CdS/TiO$_2$ 复合材料的紫外吸收增强主要归因于 ZnO 比 CdS 具有更高的电荷迁移率和更快速

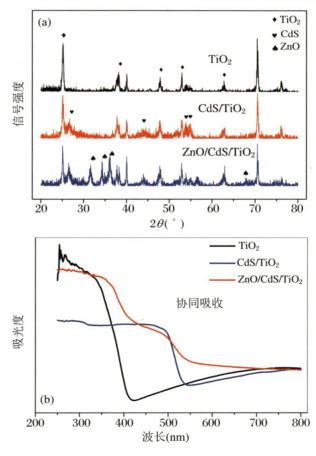

图 8.5　(a) ZnO/CdS/TiO$_2$ 的 XRD 图;(b) 紫外-可见漫反射图

的电荷分离能力[11]。而提高可见光吸收的原因可能是由于增加了光程从而提高了捕光效率、减少了反射和多次散射，形成额外的光捕捉。从几何上说，这是由于 ZnO 沉积后表面更加粗糙，表面积更大所导致的（图 8.2）[15,21]。并且，DRS 结果清楚地表明，在紫外-可见光辐照下，底层 TiO_2 未被热力学激活，但上层 ZnO 和中层 CdS 分别被紫外和可见光同时激活，这主要归因于 ZnO 薄膜的优异透明性[29,40-41]。此外，DRS 结果进一步表明了 $ZnO/CdS/TiO_2$ 异质结的独特协同机制：光催化活性最初来源于上层 ZnO 和中层 CdS，而不是底层的 TiO_2。

为了研究复合材料的光电化学性质，在太阳光照射下测量了复合材料的瞬态光电流响应和测定·OH 的相对产量。测试条件为短路光电流响应（支持电解质：30 ml 的 0.1 mol/L Na_2SO_4，阳极电位：+0.3 V/SCE，UV-vis 强度：350 W 氙气灯，有效电极面积：6.0 cm^2，pH：自然，温度：约 20 ℃，磁力搅拌速度：0 r/min，脉冲时间：100 s）和荧光光谱（电解质：30 mL 3 mmol/L 对苯二甲酸，UV-vis 强度：350 W 氙灯，有效电极面积：6.0 cm^2，pH：11.0，温度：约 20 ℃，磁力搅拌速度：500 r/min，反应时间：20 min）。结果显示 $ZnO/CdS/TiO_2$ 上的短路光电流比 CdS/TiO_2 上的短路光电流高得多，且更稳定（图 8.6(a)）。这表明，新型复合材料的协同效应使其具有优异的光催化活性和更强的抗光腐蚀能力。此外，半定量测量的·OH 也表现出类似的趋势。2-羟基对苯二甲酸与 $ZnO/CdS/TiO_2$ 光催化反应生成·OH 而产生的荧光强度比在 CdS/TiO_2 上的强得多，CdS/TiO_2 材料在 20 min 内没有任何信号（图 8.6(b)），这意味着在相同条件下新型复合材料的·OH 强度更高，显示其可以产生更多的活性氧化物种，从而提高其光催化活性（图 8.6(b)）。

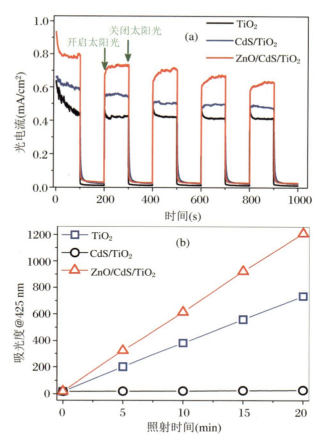

图 8.6　在太阳光照射下，TiO_2、CdS/TiO_2 和 $ZnO/CdS/TiO_2$ 的 +0.30 V 短路光电流响应（相对于 SCE）(a) 和 423.5 nm 的荧光强度 (b)

8.3 氧化锌/硫化镉/二氧化钛异质结的光催化阿特拉津降解性能及降解路径分析

为了评估复合材料的应用潜力，光催化活性通过以下实验测定：广泛应用于工业中的典型偶氮染料罗丹明 B（浓度为 5 mg/L）和广泛存在于水环境中的典型除草剂阿特拉津(浓度 5 mg/L)的降解；以及在室温下，300～770 nm 太阳光辐射，在定制的 60 mL 光反应器中处理实际纺织印染废水。光源是一个 350 W 的氙弧灯。每次实验用 30 mL 配制或实际的废水和 2 cm×3 cm 的光催化膜。辐照前，有机物在催化剂上达到吸附-脱附平衡。定期取 1 mL 样品，经 0.45 μm 滤膜过滤后再进行分析。

原纺织印染废水从合肥市一家工厂收集，储存在 4 ℃ 冰箱中，以保持其特性不变。初始 pH 范围为 6.8～7.5，离子电导率为 1.37～2.00 mS/cm，初始化学需氧量、总凯氏氮、铵态氮、硫酸盐和氯化物分别为约 7850 mg/L、1240 mg/L、920 mg/L、280 mg/L 和 610 mg/L。原料纺织废水用蒸馏水适当稀释，以获得恒定的初始浓度，即真实废水(80 mg/L TOC)，温度约 20 ℃。

首先进行了光催化罗丹明 B(RhB)降解试验。在 2 h 的模拟太阳光照射下，利用 UV-vis 吸收光谱仪对其降解过程进行测定。CdS/TiO_2 对 RhB 的去除率仅为 20% 左右，在相同条件下，$ZnO/CdS/TiO_2$ 对 RhB 的去除效率约为 90%，比 CdS/TiO_2 提高了将近 4 倍(图 8.7(a))。并且，根据图 8.7(b)计算得出，在 2 h 内，CdS/TiO_2 和 $ZnO/CdS/TiO_2$ 的每平方厘米电极表面对 RhB 的绝对去除量分别为 7.5 μg 和 30 μg(图 8.7(b))。为了避免光敏化作用的影响，更准确地评估其对环境污染物的降解能力，农药阿特拉津被用作目标污染物进行光催化降解。并且使用高效液相色谱法测定阿特拉津，采用 Hypersil-ODS 反相色谱柱，以可变波长紫外检测器(VWD)在 254 nm 处测定。流动相是水和甲醇(体积比为 40∶60)的混合物，流速为 1 mL/min。同样，在没有牺牲试剂的情况下，CdS/TiO_2 表现出较低的光催化活性(图 8.7(c))。即使加入 0.1 mol/L 的 Na_2S 和 0.01 mol/L 的 Na_2SO_3 牺牲剂后，也没有明显增加 CdS/TiO_2 光催化降解阿特拉津的效果(图 8.8(a))。与之相比，在 3 h 反应中，$ZnO/CdS/TiO_2$ 去除了 80% 以上的阿特拉津，远高于其他光催化剂的降解效率(图 8.7(c))。对于复合材料 $ZnO/CdS/TiO_2$，电极的阿特拉津去绝对除约为 25 μg/cm^2，而 ZnO/TiO_2、ZnO/CdS、TiO_2 和 ZnO 对阿特拉津的去除率分别为 14.5 μg/cm^2、13.9 μg/cm^2、14 μg/cm^2 和 10 μg/cm^2(图 8.7(d))。这些结果清楚地表明，在表面沉积 ZnO 后，形成的 $ZnO/CdS/TiO_2$ 异质结光催化活性显著提高。此外，对于纺织废水，$ZnO/CdS/TiO_2$ 复合材料的光化学脱色率和矿化效率都远高于 CdS/TiO_2 材料(图 8.7(e、f))。经过 6 h 处理，对于 $ZnO/CdS/TiO_2$ 复合材料，实际纺织废水在 475 nm 处的脱色率约为 60%，TOC 被大幅去除，去除率为 55.9%，而在相同条件下，CdS/TiO_2 处理后脱色率约为 40%，TOC 去除率为 9.9%。具体数据见表 8.1。

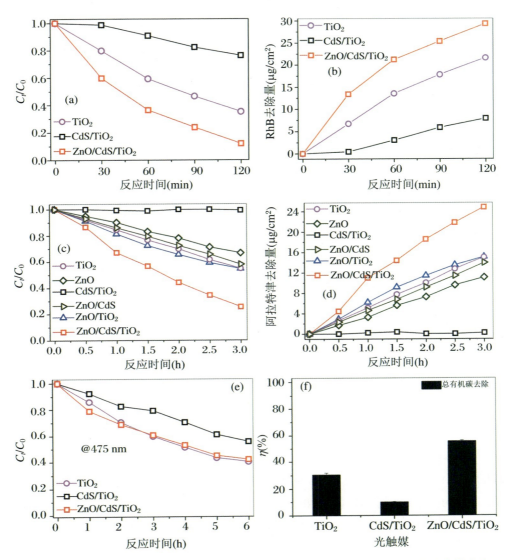

图 8.7 不同光催化剂对 RhB(a、b),阿拉特津(c、d)和纺织废水(e、f)的光催化降解效率和绝对去除量

光催化降解条件:磁力搅拌速率为 500 r/min,反应时间分别为 2 h、3 h 和 6 h

表 8.1　不同光催化剂上的污染物降解和纺织品废水处理动力学常数

光催化剂	污染物		纺织品废水	
	RhB ($\times 10^{-2}$/min)	阿拉特津 ($\times 10^{-2}$/min)	吸光度$_{475}$ ($\times 10^{-2}$/min)	TOC ($\times 10^{-2}$/min)
TiO_2	0.91	0.33	0.26	0.09
ZnO	/	0.23	/	/
CdS/TiO_2	0.23	0.002	0.16	0.02
ZnO/CdS	/	0.28	/	/
ZnO/TiO_2	/	0.34	/	/
$ZnO/CdS/TiO_2$	1.78	0.76	0.22	0.18

值得注意的是，TiO_2 在 RhB 和阿特拉津的降解以及实际纺织废水的处理方面表现出比 CdS/TiO_2 更高的活性。TiO_2 相对于 CdS/TiO_2 复合物催化效果更好可能归因于以下三个方面：独特的光源、不同的光催化氧化机制和未添加化学牺牲试剂。首先，与以往研究中在可见光源（λ>420 nm）下 CdS/TiO_2 复合物中仅激活 CdS 相比，在本研究中 CdS 和 TiO_2 在紫外-可见光（λ>300 nm）下同时激活。此外，TiO_2 的电荷迁移率和电荷分离效率明显高于 CdS[14,16,19]，从而提高了 TiO_2 的量子效率。其次，对于 CdS/TiO_2，只有电子间接氧化降解污染物，转移空穴的氧化电位降低，腐蚀性地氧化晶格硫物种和 S^{2-} [14,17]。这是因为 CdS/TiO_2 复合物中典型的Ⅱ型带隙结构能极大地降低传递电子和空穴的氧化还原电位，从而无效地释放出一部分电位能。降低的氧化还原电位可以改变光化学机制，并进一步抑制其光化学性能[37]。与此相比，由于空穴和电子同时发生间接氧化作用，未转移的空穴具有较高的氧化电位，因此 TiO_2 能协同产生大量活性氧并降解污染物。最后，本研究没有添加任何牺牲试剂，例如典型的 SO_4^{2-}/SO_3^{2-} 氧化还原介质。因此，CdS/TiO_2 在紫外光照射下在水溶液中被严重腐蚀，而在相同条件下，TiO_2 的光催化稳定性远高于 CdS/TiO_2，这也有助于提高 TiO_2 的光催化活性。

除了大大提高光催化活性外，新型 $ZnO/CdS/TiO_2$ 复合材料比传统材料如 CdS/TiO_2 具有更好的抗光腐蚀能力。在 5 次循环降解阿特拉津实验中，新型材料 $ZnO/CdS/TiO_2$ 具有良好的稳定性，阿特拉津平均去除效率为 $(80±5.0)\%$；相比之下，CdS/TiO_2 的稳定性差得多，最终去除效率仅为约 40%（图 8.8(a)）。然而，CdS/TiO_2 的光活性（48%）增加主要是由于 TiO_2 的逐渐裸露，由 TiO_2 和残留 CdS 之间的协同作用，提高了其光催化活性，并略高于单一的 TiO_2（40%）。

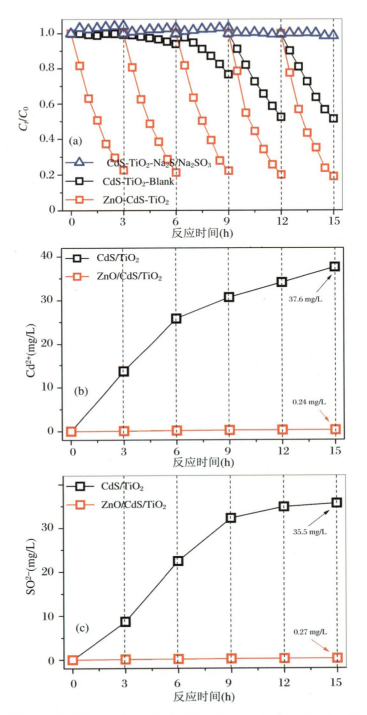

图 8.8 在相同光催化降解条件，ZnO/CdS/TiO$_2$ 和 CdS/TiO$_2$（含或不含牺牲剂(0.1 mol/L Na$_2$S + 0.01 mol/L Na$_2$SO$_3$)）对相同浓度阿特拉津的循环降解实验(a)、释放的 Cd^{2+}(b)和被氧化的 SO$_4^{2-}$ 累积浓度(c)

通过对两种 CdS 基复合物及其处理液在 5 次循环试验前后的 Cd^{2+} 和 SO_4^{2-} 含量的定量分析，进一步验证了上述结果(图 8.8(b、c))。$ZnO/CdS/TiO_2$ 复合物和 CdS/TiO_2 复合物的初始 Cd^{2+} 含量几乎相同，分别为 886.0 μg 和 900.0 μg。然而，在模拟阳光照射下 15 h 连续光催化降解后，$ZnO/CdS/TiO_2$ 复合物和 CdS/TiO_2 复合物中的 Cd^{2+} 含量分别为 845.0 μg 和 48.0 μg。经过 5 次循环光催化试验，处理后溶液中累积释放的 Cd^{2+} 含量分别为 0.24 mg/L 和 37.6 mg/L(图 8.8(b))。由于光化学强氧化的环境，从两种 CdS 基复合物中释放的 S^{2-} 主要转化为 SO_4^{2-}，并通过离子色谱测定，污染物降解完后溶液中 SO_4^{2-} 离子浓度分别为 0.27 mg/L、35.5 mg/L(图 8.8(c))。这些定量结果表明，在相同条件下，$ZnO/CdS/TiO_2$ 异质结在光化学废水处理中的结构、化学和催化稳定性远远优于 CdS/TiO_2 复合物。

为了更深入地了解两种 CdS 基催化剂的抗腐蚀性能，在阿特拉津的循环降解试验前后进一步进行了 DRS 测量图谱研究(图 8.9)。所使用的 $ZnO/CdS/TiO_2$ 在 5 次循环实验后 DRS 图谱没有表现出明显的变化，表明其具有良好的化学和光化学稳定性以及抗光腐蚀能力(图 8.9(a))；与之相对，所使用的 CdS/TiO_2 在两次循环试验后 DRS 图谱发生明显变化，且底部 TiO_2 层的光学性质明显改变，这意味着底部 TiO_2 由于上部 CdS 的化学氧化分解而裸露受到严重损害(图 8.9(b))。进一步进行了 SEM 的表征，在两次循环降解实验后，CdS/TiO_2 的形态产生了剧烈变化(图 8.10(a、b))，表明 CdS 在太阳照射下发生了严重的化学溶解，而 $ZnO/CdS/TiO_2$ 即使经过 5 次循环降解实验本质也不变(图8.10(c、d))。此外，XRD 结果表明，经过 5 次光催化试验，$ZnO/CdS/TiO_2$ 复合层中主要保留了 CdS 层(图 8.11)；而 CdS/TiO_2 复合材料中的 CdS 层，在相同条件下，仅经 2 次光催化实验后就大部分被光腐蚀。

基于此，所有抗光腐蚀表征的实验结果与 RhB 和阿特拉津的降解行为以及光电流响应的测量结果一致。这为 $ZnO/CdS/TiO_2$ 复合材料的光催化活性和抗光腐蚀能力的提高提供了更直接的证据。在这一复合材料中，上层的 ZnO 作为一种稳固、清洁的固体牺牲试剂，捕获中间层 CdS 中产生的腐蚀性的光生空穴。然而，$ZnO/CdS/TiO_2$ 复合材料向水样中释放的高毒性的 Cd^{2+} 含量(0.01～0.09 mg/L)(图 8.8)仍高于世界卫生组织(0.003 mg/L)、美国(0.005 mg/L)、欧盟(0.005 mg/L)和中国(0.005 mg/L)的饮用水等水体中 Cd^{2+} 允许值。因此，还需要进一步改进技术。

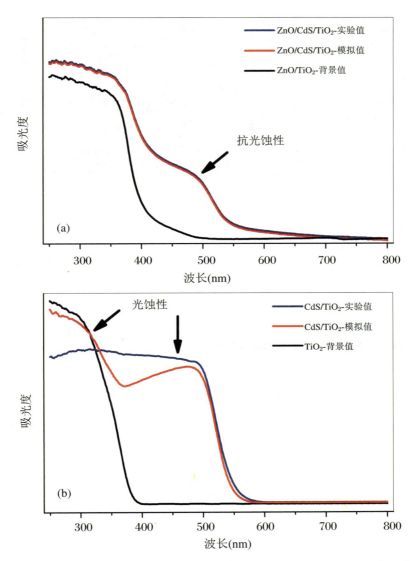

图 8.9 在相同光催化降解条件,2 次或 5 次循环使用之前和之后,ZnO/CdS/TiO_2 复合物(a)和 CdS/TiO_2 复合物(b)的紫外-可见吸收光谱的变化

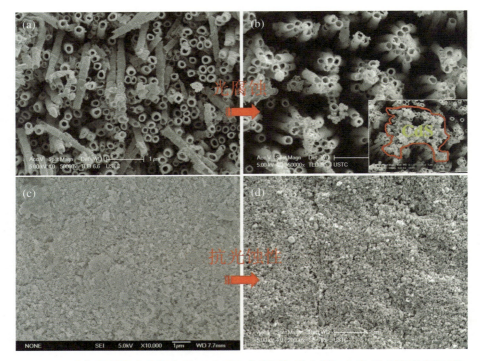

图 8.10　在相同光催化降解条件,2 次或 5 次循环使用之前和之后,ZnO/CdS/TiO$_2$ 复合物(a)和 CdS/TiO$_2$ 复合物(b)的 SEM 图

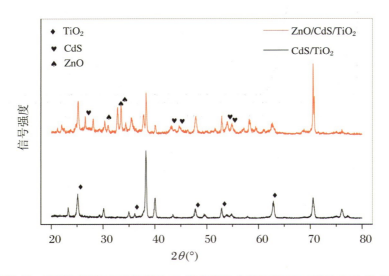

图 8.11　在相同光催化降解条件下,2 次或 5 次循环使用之前和之后,ZnO/CdS/TiO$_2$ 复合物(a)和 CdS/TiO$_2$ 复合物(b)的 XRD 图

为了进一步探索其光催化污染物的转化路径,利用液相色谱-质谱联用技术(LC-MS)对其主要降解中间体进行了鉴定。

LC-MS 系统具体参数:在 40 ℃ 条件下,采用 XTerra MS C18 液相色谱柱

(分析柱为 2.1 mm×150 mm,颗粒度为 5 μm,孔径为 136×10^{-10} m)进行分析。流速为 0.20 mL/min 的流动相以 30/70(体积比)的乙腈/水为起始流动相,在 8 min 内线性增加到 90/10,保持 2 min,最后在 1 min 内下降到 30/70 并保持 2 min。图 8.12 清晰地显示了阿特拉津在光催化过程中,·OH 作为主要的氧化活性物种,对其进行脱氯并开环,最终实现矿化。

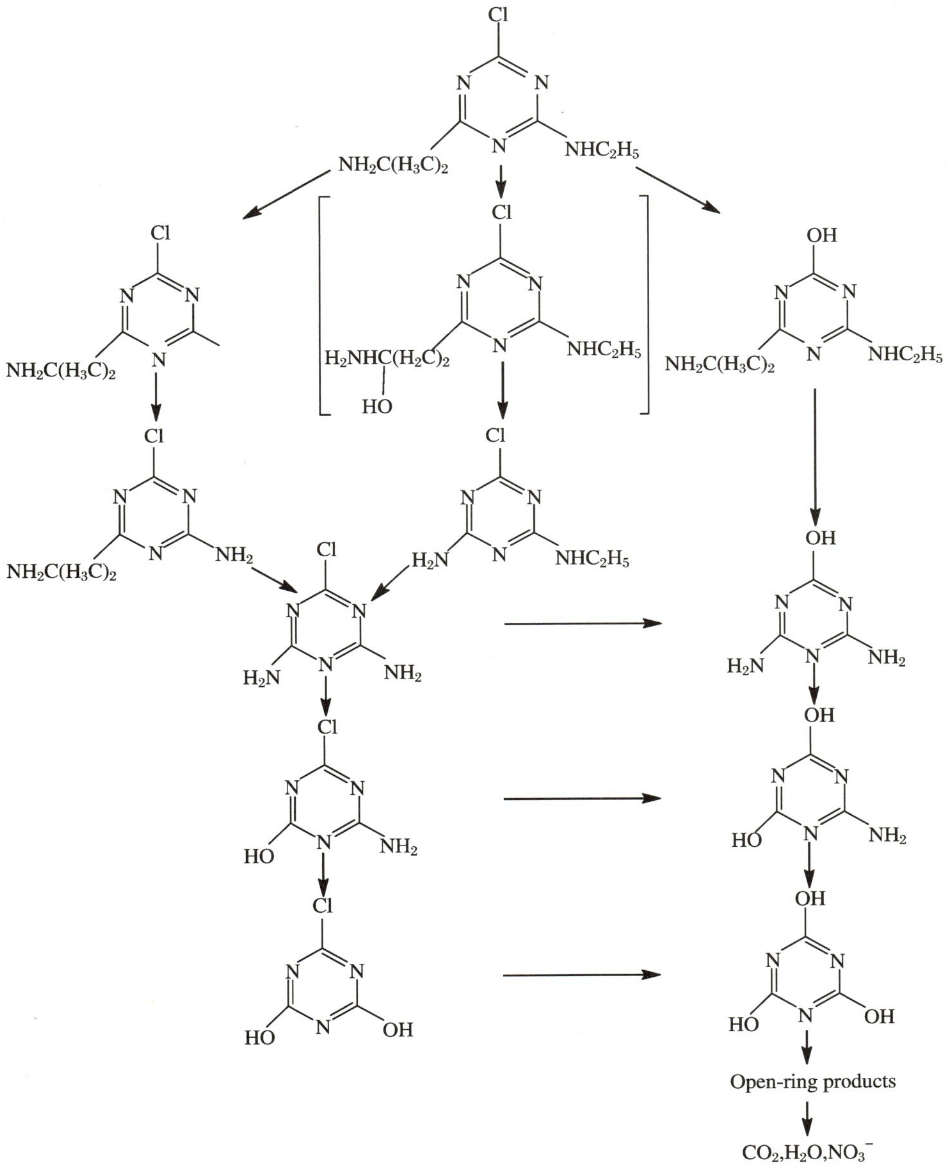

图 8.12 ZnO/CdS/TiO$_2$ 异质结光催化阿特拉津降解途径

8.4 氧化锌/硫化镉/二氧化钛异质结的光催化机制分析

当 CdS/TiO$_2$ 结合时，CdS 被光激发后，导带上的电子通过能级差，转移至 TiO$_2$ 表面，进行光催化还原，而相应的空穴保留在 CdS 表面，容易形成光腐蚀，符合实验结果。当 CdS/ZnO 结合形成著名的直接 Z 构型时，ZnO 被激发后产生的电子与 CdS 上的空穴进行了复合，减少了 CdS 的光腐蚀作用。因此，基于 ZnO 的作用，研究了 ZnO/CdS/TiO$_2$ 的工作原理（图 8.13）。首先，当 ZnO/CdS/TiO$_2$ 被模拟太阳光照射时，紫外光子被外部 ZnO 选择性吸收，而可见光子通过外部 ZnO 进入中间的 CdS 层，并完全被中间的 CdS 吸收。这种选择性吸收使外部 ZnO 被紫外光激活（反应(8.1.1)），中间 CdS 被可见光激活（反应(8.1.2)），而底部 TiO$_2$ 因为没有光子可用仍然失活。然后，ZnO 中低还原活性的导带电子可以直接与 CdS 中低氧化活性的价带空穴重新结合（反应(8.2.1)），并且 CdS 中具有较高还原活性的导带电子被转移到 TiO$_2$-NTs 表面上（反应(8.2.2)）（图8.13）。最后，ZnO 中高氧化活性的价带空穴由于具有足够高的氧化电位，可以直接或间接通过氧化 H$_2$O/OH$^-$ 从而生成·OH 来降解污染物（≡C—OH）（反应(8.3)～反应(8.5)）。同时，TiO$_2$ 中具有高还原活性的导带电子可与氧反应形成活性氧物种（反应(8.6)～反应(8.7)），它们可以间接降解污染物（反应(8.8)）。因此，直接氧化和间接氧化都协同作用于 ZnO/CdS/TiO$_2$，并显著提高了光催化活性和抗光腐蚀性能。

$$ZnO \xrightarrow{\text{紫外光}} e_{CB}^-(ZnO) + h_{VB}^+(ZnO) \tag{8.1.1}$$

$$CdS \xrightarrow{\text{可见光}} e_{CB}^-(CdS) + h_{VB}^+(CdS) \tag{8.1.2}$$

$$h_{VB}^+(CdS) + e_{CB}^-(ZnO) \rightarrow CdS + ZnO + heat \tag{8.2.1}$$

$$e_{CB}^-(CdS) + TiO_2 \rightarrow e_{CB}^-(TiO_2) + CdS \tag{8.2.2}$$

$$h_{VB}^+(ZnO) + OH^- \rightarrow \cdot OH(ZnO) \tag{8.3.1}$$

$$h_{VB}^+(ZnO) + H_2O \rightarrow \cdot OH(ZnO) + H^+ + e^- \tag{8.3.2}$$

$$\equiv C—OH + h_{VB}^+(ZnO) \rightarrow \equiv C—O^* + H^+ + e^- \rightarrow \cdots \rightarrow CO_2 + H_2O \tag{8.4}$$

$$\equiv C—OH + HO^*(ZnO) \rightarrow \equiv C—O^* + H_2O \rightarrow \cdots \rightarrow CO_2 + H_2O \tag{8.5}$$

$$e_{CB}^-(TiO_2) + O_{2ads} \rightarrow {}^*O_2^-(TiO_2) \tag{8.6.1}$$

$$2e_{CB}^-(TiO_2) + O_2 + 2H^+ \rightarrow H_2O_2 \quad (8.6.2)$$

$$H_2O_2 + UV\ light \rightarrow 2HO_{free}^* \quad (8.7.1)$$

$$H_2O_2 + e_{CB}^-(TiO_2) \rightarrow \cdot OH(TiO_2) + OH^- \quad (8.7.2)$$

$$H_2O_2 + {}^*O_2^-(TiO_2) \rightarrow \cdot OH(TiO_2) + OH^- + O_2 \quad (8.7.3)$$

$$\equiv C-OH + HO^*(TiO_2) \rightarrow \equiv C-O^* + H_2O \rightarrow \cdots \rightarrow \cdots \rightarrow CO_2 + H_2O \quad (8.8)$$

$$(e_{CB}^- + h_{VB}^+)(CdS) + TiO_2 \rightarrow h_{VB}^+(CdS) + e_{CB}^-(TiO_2) \quad (8.9)$$

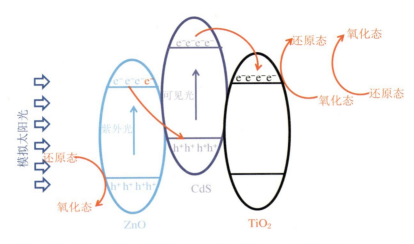

图 8.13　ZnO/CdS/TiO$_2$ 在模拟太阳光照射下的电子-空穴转移原理

相比之下，对于 CdS/TiO$_2$ 材料，在太阳照射下，仅在 CdS 内形成电子-空穴对（反应(8.1)和反应(8.2)），然后将电子转移到 TiO$_2$ 的导带上，在 CdS 的表面和/或价带上留下空穴（反应(8.9)）。然后，TiO$_2$ 中具有高还原活性的导带电子可以与氧反应形成活性氧物种（反应(8.6)和反应(8.7)），间接使污染物降解（反应(8.8)）。然而，CdS 中具有较低的氧化活性的价带空穴不能通过 H$_2$O/OH$^-$ 氧化形成的·OH 直接和间接地分解污染物，这是由于 CdS 中的价带空穴的氧化电位(+1.2~1.5 eV)比·OH/H$_2$O(+2.27 eV)的低[39,42-43]，因此不能与 H$_2$O/OH$^-$ 热力学反应产生·OH。价带空穴可以很容易地被牺牲试剂捕获，或者在没有清除剂的情况下腐蚀 CdS。因此，CdS/TiO$_2$ 上只发生间接氧化，导致其光催化活性和抗光腐蚀能力相对较差。

根据参考文献与实验结果，提出了相应较为合理的电子-空穴转移路径。电子从中间层 CdS 转移到底层的 TiO$_2$ 有两个原因：其一，与上层 ZnO 相比，底部 TiO$_2$ 有更多的正电导带，可能在热力学上使中间 CdS 中的电子更容易转移到底部 TiO$_2$ 中[29]。其二，在 ZnO/CdS/TiO$_2$ 异质结中，底部 TiO$_2$ 纳米管与中间 CdS 纳米粒子之间的接触比上层 ZnO 纳米粒子与中间 CdS 纳米粒子之间的接触更密切，使得中间的 CdS 纳米颗粒中的电子更容易转移到底部的 TiO$_2$ 纳米管中。

并且,从 CdS 到 ZnO 的电子转移效率远低于从 CdS 到 TiO$_2$ 的电子转移效率[44]。

此外,外层 ZnO 中的导带电子被转移到中间 CdS 有以下两个原因:其一,结果表明,在 5 次循环降解实验中,ZnO/CdS/TiO$_2$ 复合材料的中间层 CdS 具有优良的抗光腐蚀性能和良好的结构稳定性、化学稳定性和催化稳定性[11,13,17,19]。其二,通过直接 Z 型电荷转移机制,被激活 ZnO 的导带产生的还原性导带电子可以直接或原位消耗 CdS 价带中的光生空穴[29,40-41]。

在本章中,为了提高已知的 CdS/TiO$_2$ 复合材料在难降解污染物降解中的应用潜力,我们提出了直接利用 ZnO 中低还原性的导带电子作为固体牺牲试剂,有效地从 CdS 中捕获低氧化能的价带空穴。实验结果证明,制备的 ZnO/CdS/TiO$_2$ 复合材料在在太阳光照射下,无需任何额外的牺牲试剂,对目标污染物 RhB 和阿特拉津的降解以及纺织废水处理时,比典型的 CdS/TiO$_2$ 类似物具有更高的光催化活性。此外,在模拟太阳光照射下,由于 ZnO 的负载,使 ZnO/CdS/TiO$_2$ 复合材料在水溶液光催化反应中的光腐蚀大大降低,表观抗光腐蚀能力明显提高。由于这种新型的 ZnO/CdS/TiO$_2$ 异质结的构筑,不仅提高了催化活性,还解决了 CdS 的稳定性问题,为容易光腐蚀催化材料的设计与应用提供了一种新的解决思路,并增加了其实际应用的可能性。

参考文献

[1] CHEN Q, WANG H, WANG C, et al. Activation of molecular oxygen in selectively photocatalytic organic conversion upon defective TiO$_2$ nanosheets with boosted separation of charge carriers[J]. Appl. Catal. B, 2019, 262:118258.

[2] CADIAU A, KOLOBOV N, SRINIVASAN S, et al. A titanium metal-organic framework with visible-light-responsive photocatalytic activity[J]. Angew. Chem. Int. Ed., 2020, 59: 13468-13472.

[3] FANG Y, LIU Z, HAN J, et al. High performance electrocatalytic conversion of N$_2$ to NH$_3$ using xxygen acancy in ich TiO$_2$ in situ grown on Ti$_3$C$_2$Tx mxene[J]. Adv. Energy Mater., 2019, 9: 1803406.

[4] LIN J, LI P, XU H, et al. Controlled synthesis of mesoporous single-crystalline TiO$_2$ nanoparticles for efficient photocatalytic H$_2$ evolution[J]. J. Hazard. Mater., 2020, 391:122530.

[5] HAN S, NIU Q, QIN N, et al. In situ growth of M-{001}TiO$_2$/Ti photoelectrodes: synergetic dominant {001} facets and ratio-optimal surface junctions for the effective oxidation of environmental pollutants[J]. Chem. Comm., 2020, 56. DOI: 10.1039/C9CC0929J.

[6] CHENG X, DONG G, ZHANG Y, et al. Dual-bonding interactions between MnO$_2$ cocatalyst and TiO$_2$ photoanodes for efficient solar water splitting[J]. Appl. Catal. B, 2020, 267: 118723.

[7] MENG N C, JIN B, CHOW C, et al. Recent developments in photocatalytic water treatment technology: a review[J]. Water Res., 2010, 44: 2997-3027.

[8] HERNANDEZ-ALONSO M, FRESNO F, SUAREZ S, et al. Development of alternative photocatalysts to TiO$_2$: challenges and opportunities[J]. Energy Environ. Sci., 2009, 2: 1231-1257.

[9] ZHANG J, TANG B, ZHU Z, et al. Size-controlled microporous SiO$_2$ coated TiO$_2$ nanotube arrays for preferential photoelectrocatalytic xxidation of highly toxic PAEs[J]. Appl. Catal. B-Environ., 2019, 268: 118400.

[10] IAN X, BRENNECKA G L, TAN X. Direct observations of field-intensity-dependent dielectric breakdown mechanisms in TiO$_2$ single nanocrystals[J]. ACS Nano, 2020, 14: 8328-8334.

[11] SUN W T, YU A, PAN H Y, et al. CdS quantum dots sensitized TiO$_2$ nanotube-array photoelectrodes[J]. J. Am. Chem. Soc., 2009, 130: 1124-1125.

[12] LOEB S K, ALVAREZ P, BRAME J A, et al. The technology horizon for photocatalytic water treatment: sunrise or sunset?[J]. Environ. Sci. Technol., 2018. 53: 2937-2947

[13] SØNDERGAARD-PEDERSEN F, BROGE N, YU J, et al. Maximizing the catalytically active {001} facets on anatase nanoparticles[J]. Chem. Mater., 2020, 32: 5134-5141.

[14] DASKALAKI V M, ANTONIADOU M, PUMA G L, et al. Solar light-responsive Pt/CdS/TiO$_2$ photocatalysts for hydrogen production and simultaneous degradation of inorganic or organic sacrificial agents in wastewater[J]. Environ. Sci. Technol., 2010, 44: 7200-7205.

[15] GAO X F, SUN W T, HU Z D, et al. An efficient method to form heterojunction CdS/TiO$_2$ photoelectrodes using highly ordered TiO$_2$ nanotube array films[J]. J. Phys. Chem. C, 2009, 113: 20481-20485.

[16] LEE Y L, CHANG C-H. Efficient polysulfide electrolyte for CdS quantum dot-sensitized solar cells[J]. J. Power Sources, 2008, 185: 584-588.

[17] LI G S, ZHANG D Q, YU J C. A new visible-light photocatalyst: CdS quantum dots embedded mesoporous TiO_2[J]. Environ. Sci. Technol., 2009, 43: 7079.

[18] LI Q, GUO B, YU J, et al. Highly efficient visible-light-driven photocatalytic hydrogen production of CdS-cluster-decorated graphene nanosheets[J]. J. Am. Chem. Soc., 2011, 133: 10878-10884.

[19] TADA H, MITSUI T, KIYONAGA T, et al. All-solid-state Z-scheme in CdS-Au-TiO_2 three-component nanojunction system[J]. Nat. Mater., 2006, 5: 782-786.

[20] LOW J, DAI B, TONG T, et al. In situ irradiated X-ray photoelectron spectroscopy investigation on a Direct Z-scheme TiO_2/CdS composite film photocatalyst[J]. Adv. Mater., 2019, 31: 1802981.

[21] YANG W Z X L H L D T J, PENG J. Coaxial heterogeneous structure of TiO_2 nanotube arrays with CdS as a superthin coating synthesized via modified electrochemical atomic layer deposition[J]. J. Am. Chem. Soc., 2010, 132: 12619-12626.

[22] KRIPAL R, VAISH G, TRIPATHI U M. Comparative study of EPR and optical properties of CdS, TiO_2 and CdS-TiO_2 nanocomposite[J]. J. Electron. Mater., 2019, 48: 1545-1552.

[23] WU K, WU P, ZHU J, et al. Synthesis of hollow core-shell CdS@TiO_2/Ni_2P photocatalyst for enhancing hydrogen evolution and degradation of MB[J]. Chem. Eng. J., 2019, 360: 221-230.

[24] MEISSNER D, MEMMING R, KASTENING B. Photoelectrochemistry of cadmium sulfide. 1. reanalysis of photocorrosion and flat-band potential[J]. J. Phys. Chem., 1988, 92. DOI: 10.1021/j100323a032.

[25] REBER J F, RUSEK M. ChemInform abstract: Photochemical hydrogen production with platinized suspensions of cadmium sulfide and cadmium zinc sulfide modified by silver sulfide[J]. Cheminform, 1986, 17. DOI: 10.1002/chin.198623026.

[26] SAPP S A, ELLIOTT C M, CONTADO C, et al. Substituted polypyridine complexes of cobalt(Ⅱ/Ⅲ) as efficient electron-transfer mediators in dye-sensitized solar cells[J]. J. Am. Chem. Soc., 2002, 124: 11215-11222.

[27] TIAN Y, TATSUMA T. Mechanisms and applications of plasmon-induced charge separation at TiO_2 films loaded with gold nanoparticles[J]. J. Am. Chem. Soc., 2005, 127:7632-7637.

[28] SHEENEY-HAJ-ICHIA L, POGORELOVA S, GOFER Y, et al. Enhanced photoelectrochemistry in CdS/Au nanoparticle bilayers[J]. Funct. Mater., 2004, 14: 416-424.

[29] WANG Z-S, HUANG C-H, HUANG Y-Y, et al. A highly efficient solar cell made from a dye-modified ZnO-covered TiO_2 nanoporous electrode[J]. Chem. Mater., 2001, 13:678-682.

[30] CAZZANTI S, CARAMORI S, ARGAZZI R, et al. Efficient non-corrosive electron-transfer mediator mixtures for dye-sensitized solar cells [J]. J. Am. Chem. Soc., 2006, 128:9996-9997.

[31] JANG J S, KIM H G, BORSE P H, et al. Simultaneous hydrogen production and decomposition of H_2S dissolved in alkaline water over CdS-TiO_2 composite photocatalysts under visible light irradiation[J]. Int. J. Hydrog. Energy, 2007, 32:4786-4791.

[32] DIBY N D D, DUAN Y, GRAH P A, et al. Enhanced photoelectrochemical water-splitting performance of TiO_2 nanorods sensitized with CdS via hydrothermal approach[J]. J. Alloys Compd., 2019, 803:456-465.

[33] CHEN Q, LI J, LI X, et al. Visible-light responsive photocatalytic fuel cell based on WO_3/W photoanode and Cu_2O/Cu photocathode for simultaneous wastewater treatment and electricity generation[J]. Environ. Sci. Technol., 2012, 46:11451-11458.

[34] SAYAMA K, MUKASA K, ABE R, et al. A new photocatalytic water splitting system under visible light irradiation mimicking a Z-scheme mechanism in photosynthesis[J]. Chemistry J. Photoch. Photobio. A, 2002, 148: 71-77.

[35] LIU Y, TIAN J, WEI L, et al. A novel microwave-assisted impregnation method with water as the dispersion medium to synthesize modified g-C_3N_4/TiO_2 heterojunction photocatalysts[J]. Opt. Mater., 2020, 107:110128.

[36] LIU Y, TIAN J, WEI L, et al. Modified g-C_3N_4/TiO_2/CdS ternary heterojunction nanocomposite as highly visible light active photocatalyst originated from CdS as the electron source of TiO_2 to accelerate Z-type heterojunction [J]. Sep. Purif. Technol., 2020, 257. DOI: 10.1016/j. seppur.

2020.117976.

[37] LIU G, WANG L, YANG H G, et al. Titania-based photocatalysts-crystal growth, doping and heterostructuring[J]. J. Mater. Chem., 2010, 2: 831-843.

[38] SAYAMA K, MUKASA K, ABE R, et al. Stoichiometric water splitting into H_2 and O_2 using a mixture of two different photocatalysts and an IO_3-I- shuttle redox mediator under visible light irradiation[J]. Chem. Comm., 2001: 2416-2417.

[39] WANG G L, XU J J, CHEN H Y, et al. Label-free photoelectrochemical immunoassay for α-fetoprotein detection based on TiO_2/CdS hybrid[J]. Biosens. Bioelectron., 2009, 25:791-796.

[40] WANG X, LIU G, LU G Q, et al. Stable photocatalytic hydrogen evolution from water over ZnO-CdS core-shell nanorods[J]. Int. J. Hydrog. Energy, 2010, 35:8199-8205.

[41] WANG X, YIN L, LIU G, et al. Polar interface-induced improvement in high photocatalytic hydrogen evolution over ZnO-CdS heterostructures[J]. Energy Environ. Sci., 2011, 4:3976-3979.

[42] LIN C J, YU Y H, LIOU Y H. Free-standing TiO_2 nanotube array films sensitized with CdS as highly active solar light-driven photocatalysts[J]. Appl. Catal. B-Environ, 2009, 93:119-125.

[43] WU L, YU J C, FU X. Characterization and photocatalytic mechanism of nanosized CdS coupled TiO_2 nanocrystals under visible light irradiation[J]. J. Mol. Catal. A-Chem., 2006, 244:25-32.

[44] SPANHEL L, WELLER H, HENGLEIN A. Photochemistry of semiconductor colloids. 22. electron ejection from illuminated cadmium sulfide into attached titanium and zinc oxide particles[J]. J. Am. Chem. Soc., 1987, 109:6632-6635.

第 9 章

二维二氧化钛/氮化碳异质结的构筑及其光电催化性能

9.1 二氧化钛与非金属半导体异质结的研究进展

正如前文所述,合理的构筑异质结材料会暴露的更多的活性位点,这对其催化性能起着重要作用。同样,二氧化钛宽的带隙(3.2 eV)、有限的暴露活性位点和一般的导电性严重限制了其在光电催化反应中的应用[1-2]。迄今为止,人们已经采取了各种方法进行光电催化材料的优化,如掺杂、构筑异质结、复合助催化剂和调控晶面等[3-8]。通过调研发现,已有文献报道制备了具有高活性面的二氧化钛电极,用于光电催化反应中的污染物降解[6,9-11]。有研究通过碳掺杂、氮掺杂或自掺杂成功地制备了 TiO_2 纳米管阵列光电极,用来光电催化降解有机污染物[6,12-13]。此外,也有文献报道使用 TiO_2/Ti 旋转圆盘光电催化反应器处理 RhB 溶液和工业纺织污水获得优异的光电催化性能[14-16]。有研究为了提高二氧化钛电极的效率、稳定性和优良的催化性能,开发了新型金属/金属硫化物助催化剂改性异质结[17-19]。虽然人们已经对光电催化反应的性能提高进行了一些研究,但建立一种简单、低成本的方法来构建具有优良电化学和光化学特性的高性能二氧化钛电极也是非常有价值的。因此,异质结的构建策略可以为解决这些难题提供一种有效的方法,是改善电子-空穴有效分离和转移的一种有前景的策略。

考虑到这一问题,$g-C_3N_4$ 具有类石墨烯的层状结构,可作为提高可见活性和导电性的预期材料。$g-C_3N_4$ 作为一种典型的无金属光催化剂,由于其优异的热稳定性和化学稳定性、相对廉价和无毒性,引起了人们的极大关注[20-21]。基于 $g-C_3N_4$ 光催化的光电化学电池已被成功应用于 O_2/H_2O 氧化还原偶合来产生 H_2O_2[22-23]。此外,也有文献报道,通过负载 $g-C_3N_4$,所获得材料增强了可见光吸收,提高光电化学性能,被用于敏感检测和水处理[23-24]。近年来,不同种类的 TiO_2-$g-C_3N_4$ 材料也被成功合成,实现有效电荷分离,用于光催化污染物降解,如中空 TiO_2 球体/$g-C_3N_4$、纳米颗粒 TiO_2-$g-C_3N_4$ 和介孔 TiO_2-$g-C_3N_4$ 等[25-31]。然而,在光电催化性能方面,TiO_2-$g-C_3N_4$ 不仅需要高效的光催化活性而且需要和阳极材料之间具有良好的导电性[31-32]。因此,根据 DFT 计算结果,为改善阳极表面的电子转移,研究光电催化应用中的协同效应,需要设计合成具有有效电子

转移路径的二维 TiO_2-g-C_3N_4。

本工作成功地合成了 Ti—N 和 C—O 键连接的二维 TiO_2-g-C_3N_4，并将其与碳纸组装成电极，用于光电催化污染物降解。接着，对二维 TiO_2-g-C_3N_4 的光学性质、光致电荷的分离和转移以及电化学性质进行了系统的表征，并且分析了其光电催化过程中的活性氧种类。然后，通过密度泛函(DFT)计算，系统地探讨了二维 TiO_2-g-C_3N_4 材料中的界面相互作用。此外，还对二维 TiO_2-g-C_3N_4/碳纸电极降解 BPA 的性能、稳定性和机制进行了评价。最后，研究了二维 TiO_2-g-C_3N_4 在光电催化反应中可能的协同效应。这项研究为高效二氧化钛基光电子催化剂的设计提供了新的见解，可用于环境中污染物降解光电催化。

9.2 二维二氧化钛与氮化碳复合材料的合成

9.2.1 二维二氧化钛与氮化碳复合材料的合成及电极制备

根据先前的文献方法，我们合成二维二氧化钛纳米片。首先，配制溶液 A，将 0.048 g P123(PEO20-PPO70-PEO20，平均分子量～5800，Sigma-Aldrich) 和 7 mL 乙醇加入烧杯，接着搅拌 30 min 以获得透明溶液，备用。同理，用 1 mL 钛酸四丁酯(TBOT) 和 1 mL 浓盐酸加入烧杯，搅拌 20 min，得到透明溶液，记为溶液 B。接着，将溶液 A 逐滴添加到溶液 B 中，随后添加 30 mL 乙二醇(EG)。在剧烈搅拌 30 min 后，将所得黄色透明溶液移入聚四氟乙烯内衬的不锈钢高压釜中(体积为 50 mL)，并将其在 160 ℃下加热 12 h。当高压釜冷却至室温时，取出样品用蒸馏水和乙醇充分超声清洗。最后，将样品在空气中煅烧(400 ℃,2 h)。

紧接着，将先前制备的介孔二氧化钛充分研磨，并取 60 mg 加入 15 mL 水中,在剧烈搅拌下加入 5 g 尿素溶解。搅拌 2 h 后,将溶液移入坩埚中并在烘箱(60 ℃)中保存过夜。最后,将所得前驱体在空气中 500 ℃下煅烧 2 h(加热速率为 5 ℃/min)。反应完成并冷却至室温后,收集样品用于电极制备。实验中使用的试剂纯度均为分析纯,且未进一步处理。

首先分别将 5 mg 二维 TiO_2-g-C_3N_4、TiO_2、g-C_3N_4 和商品化 P25 分别分散在 2.5 mL 去离子水中。接着,通过超声分散,用移液枪滴涂到碳纸电极上(0.5 mL)。将所得碳纸电极在空气中 450 ℃煅烧 1 h。煅烧完成后,收集制备的电极,准备用于随后的降解实验。

9.3 二维二氧化钛与氮化碳复合材料电荷密度散度的 DFT 计算

对于该材料,我们首先利用密度泛函理论(DFT)完成该模型的理论计算预测,该理论来源于 Perdew-Burke-Ernzerhof(PBE)交换相关泛函的广义梯度近似。利用 VASP 软件进行计算,该软件包以平面波为基础跨越倒易空间,并使用投影增强波方法。对于平面波基集,使用 $Ecut$ = 500 eV 的截止值。将含 12 个 C 原子和 16 个 N 原子的 1×1 单层 g-C_3N_4 放置在含 24 个 Ti 原子和 48 个 O 原子的 1×2 TiO_2{001} 表面板上。为了分离周期图像之间的相互作用,在平板上方放置约 $15×10^{-10}$ m 的真空度。在弛豫过程中,允许结构的原子在晶格参数为 $7.37(10^{-10}$ m$)×11.877(10^{-10}$ m$)×27.284(10^{-10}$ m$)$ 的超电池中以 $1×10^{-4}$ eV 的能量收敛和 $0.03×10^{10}$ eV/m 的强制收敛进行调节。采用 2×3×1 的 Monkhorst 块 K 点网格对二维 Brillouin 区域进行采样。

在确定模型及参数计算后,为了评估制备的二维 TiO_2-g-C_3N_4 异质结上的电荷分离和输运,分析了二维 TiO_2-g-C_3N_4 异质结构的电荷密度发散,结果如图 9.1(a)和(b)所示(青色区域表示电荷耗尽,黄色区域表示电荷积聚)。理论计算显示电荷重分布主要发生在二维 TiO_2-g-C_3N_4 界面区,而在远离界面的 TiO_2 上几乎没有观察到电荷变化,这主要是由于 TiO_2 和 g-C_3N_4 之间的弱范德华力相互作用[33-34]。二维 TiO_2-g-C_3N_4 异质结的电荷分布类似于 n-n 结的电荷分布[34]。平面平均电荷密度差显示沿 Z 方向电荷密度的变化（如图 9.1(b)所示）。正数表示电子积聚，负数表示电子耗尽。结果表明，界面处的变化表明电子从界面上的 g-C_3N_4 移动到 TiO_2，而空穴仍留在 g-C_3N_4 侧。净电荷积累有助于在界面处形成一个内建电场，这有利于电子-空穴对的分离[34-35]。这些理论电化学优势预测，二维 TiO_2-g-C_3N_4 是一种优良的 PEC 降解有机污染物的材料。因此，应构建二维 TiO_2-g-C_3N_4 复合材料的二维结构，以便在 PEC 性能中暴露活性位点并促进其界面的电子转移。

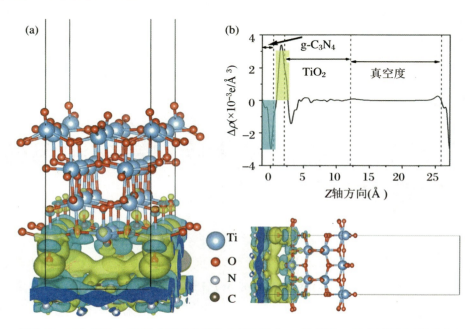

图 9.1　二维 TiO_2-g-C_3N_4 电荷密度差的侧视图(a)；二维 TiO_2-g-C_3N_4 的平面平均电子密度差 $\Delta\rho(z)$ (b)。青色和黄色区域分别表示电子消耗和积累

9.4
二维二氧化钛与氮化碳复合材料的表征

9.4.1
二维二氧化钛与氮化碳复合材料的微观结构分析

基于 DFT 计算,设计了二维 TiO_2-g-C_3N_4 样品的合成路线,如图 9.2(a)所示。根据示意图,使用改进的方法通过水热反应合成了 TiO_2。然后,通过煅烧反应合成了 g-C_3N_4 和二维 TiO_2-g-C_3N_4 材料。获得的样品的微观形貌及元素都在蔡司扫描电子显微镜上测定。同理,样品的高分辨率透射电子显微镜(HRTEM)图像由扫描透射电子显微镜测定(扫描透射电镜,JSM-6330FT,JEOL公司,日本)。图 9.3 显示获得的 TiO_2 样品均匀、分散,主体是由纳米粒子组成的纳米片构成(制备的 g-C_3N_4 样品如图 9.4 所示,符合文献的报道为片状结构)。此外,图 9.2(b~d)中提供了使用所获得的 TiO_2 样品和尿素混合通过煅烧方法合成的二维 TiO_2-g-C_3N_4 样品的 SEM 图像。结果显示,所获得的二维 TiO_2-g-C_3N_4 样品是均匀的,并且为二维结构(图 9.2(c))。进一步分析了合成的二维 TiO_2-g-C_3N_4 样品元素分布,EDS 测试直接证实 TiO_2 在 g-C_3N_4 相表面的较好分散性(图 9.2(d))。此外,TEM(图 9.5)和 AFM(图 9.6)图像又共同证实了 TiO_2-g-C_3N_4 样品的二维结构。为了直观证明当样品负载到碳纸电极后的形貌变化,通过制备电极的 SEM 图像表征,当 TiO_2-g-C_3N_4 样品涂到碳纸电极表面后形貌并没有改变,同时也未改变碳纸的形貌(图 9.7)。

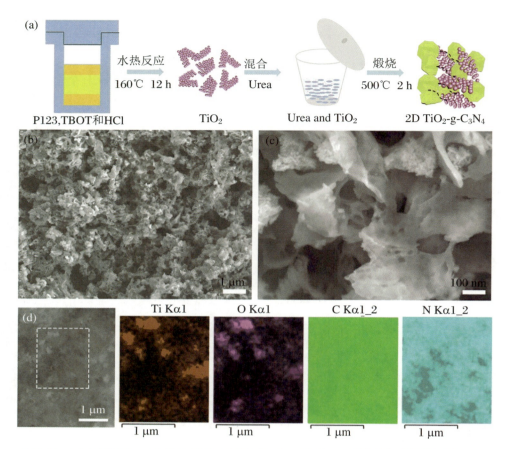

图 9.2 (a) 二维 TiO$_2$-g-C$_3$N$_4$ 的制备示意图;(b、c) 二维 TiO$_2$-g-C$_3$N$_4$ 的扫描电镜图像;(d) 二维 TiO$_2$-g-C$_3$N$_4$ 的扫描电镜图像中的选定区域和 EDS 映射

图 9.3 TiO$_2$ 的 SEM 图像

图 9.4　g-C$_3$N$_4$ 的 SEM 图像

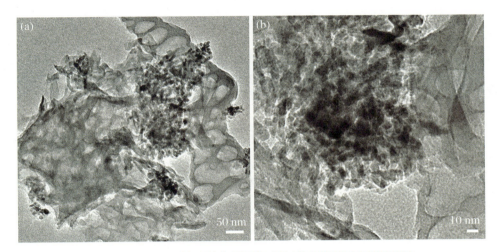

图 9.5　(a) 二维 TiO$_2$-g-C$_3$N$_4$ 的透射电镜图像；(b) 高分辨率透射电镜图像

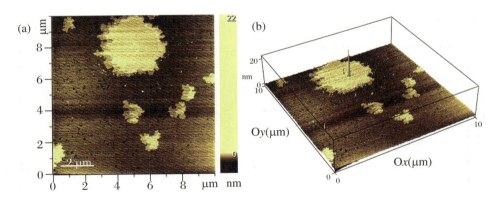

图 9.6　二维 TiO$_2$-g-C$_3$N$_4$ 的 AFM 图像

图 9.7　CF 电极表面涂层的二维 TiO$_2$-g-C$_3$N$_4$ 的扫描电镜图像

9.4.2
二维二氧化钛与氮化碳复合材料的光物理性质及光化学性质

除了扫描电镜(SEM)和能谱仪(EDS)图谱分析外,还利用粉末 X 射线衍射仪(XRD,型号 RINT-2500,日本日经株式会社;使用铜 Kα 辐射源(λ = 0.1541841 nm),以 8°/min 的扫描速率)对 TiO$_2$、g-C$_3$N$_4$ 和二维 TiO$_2$-g-C$_3$N$_4$ 进行物相测定,使用傅立叶变换红外光谱法(FTIR,Vertex70,Bruker 公司,德国)对所得材料官能团进行了分析。所得材料的 XRD 图展示于图 9.8(a)中。二维 TiO$_2$-g-C$_3$N$_4$ 样品具有两种特征峰,即分别与 TiO$_2$(标准锐钛矿 JCPDS 编号

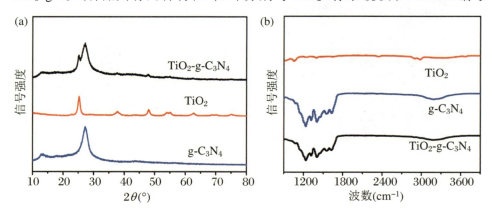

图 9.8　(a) 获得的 TiO$_2$、g-C$_3$N$_4$ 和二维 TiO$_2$-g-C$_3$N$_4$ 的 XRD 图谱;(b) 获得的 TiO$_2$、g-C$_3$N$_4$ 和二维 TiO$_2$-g-C$_3$N$_4$ 的 FT-IR 光谱

21-1272)和 g-C_3N_4 一致，表明该材料已成功合成。对制备样品进行 FTIR 光谱分析，以确认二维 TiO_2-g-C_3N_4 中存在 g-C_3N_4（图 9.8(b)）[25-27]。二维 TiO_2-g-C_3N_4 样品在 1200～1700 cm^{-1} 处有明显的峰，这与 g-C_3N_4 特征的芳香碳氮杂环拉伸震动一致[36]。

此外，通过 X 射线光电子能谱（XPS, phi 5600, Ulvac Phi 公司，日本）测定和分析了 TiO_2、g-C_3N_4 和二维 TiO_2-g-C_3N_4 的组成及其化学状态。通过 XPS 全谱，发现复合材料含有 C 1s、N 1s、Ti 2p 和 O 1s 峰（图 9.9(a)），这与 FTIR 光谱和 EDS 图谱的结果一致。为了深入探索该合成材料的化学状态，测量了高分辨率的 XPS 光谱（图 9.9(b~d)）。从 Ti 2p XPS 光谱（图 9.9(b)）可以明显看出：① Ti $2p_{1/2}$ 到 Ti $2p_{3/2}$ 信号峰的结合能（5.92 eV）与先前文献一致[36-38]；② 二维 TiO_2-g-C_3N_4 样品的 Ti 2p 光谱在 C 和/或 N 掺杂与纯二氧化钛相同。此外，与获得的 g-C_3N_4 相比，二维 TiO_2-g-C_3N_4 样品的 C 1s 和 N 1s 光谱也分别显示了一个新的峰（图 9.9(c、d)）。N 1s 峰值 397.5 eV 应归因于 Ti—N，而相应的 C 1s 峰值 287.8 eV 应归因于 C═O[36]。因此，这些研究充分证实，掺杂的 C 和 N 原子明显地与 TiO_2 结合，这增强了 TiO_2 和 g-C_3N_4 之间的电子-空穴对分离[37]。

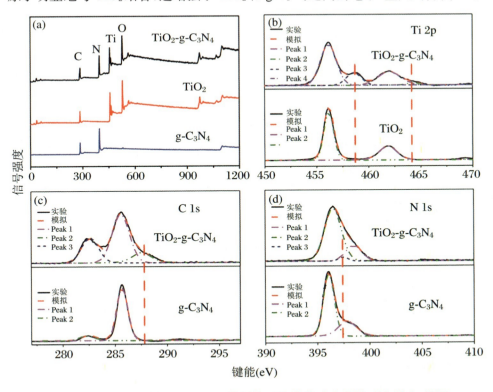

图 9.9 TiO_2、g-C_3N_4 和二维 TiO_2-g-C_3N_4 样品的 XPS 光谱：(a) 全谱；(b) Ti 2p 谱图；(c) C 1s 谱图；(d) N 1s 谱图

9.4.3
TiO₂/g-C₃N₄材料的光物理和光电化学性质

电子结构对半导体材料的光学吸收及其催化性能有很大影响。通过所得样品的紫外-可见漫反射光谱（紫外-可见分光光度计，UV-3100，日本岛津公司）证实，合成的二维 TiO₂-g-C₃N₄ 样品的吸收边缘位置显示出明显的红移（图 9.10 (a)）。进一步，根据先前文献的计算公式[39-40]，可以计算推出 TiO₂-g-C₃N₄、TiO₂、g-C₃N₄ 和 P25 半导体的带隙，分别约为 2.67 eV、3.11 eV、2.67 eV 和 2.97 eV（如图 9.11 所示）。为了进一步研究光电转换的性能，所有的光电化学测量都在一个电化学工作站上进行（CHI 660d，中国辰华公司）。电化学阻抗谱分析在 1 mol/L Na₂SO₄ 水溶液中进行，频率范围为 10^5 至 10^{-2} Hz。在这些电化学测量过程前，玻碳电极（GC）需要用氧化铝粉末（0.05 μm）抛光，然后在 Milli-Q 水中超声清洗，每次样品定量滴定。此外，铂丝和 Ag/AgCl 电极分别用作对电

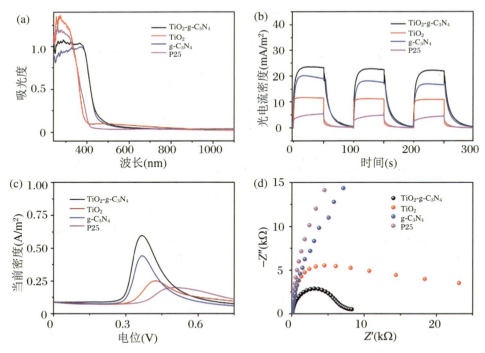

图 9.10 (a)紫外-可见漫反射光谱；(b) 瞬态光电流密度曲线；(c) 电化学活性的微分脉冲阳极溶出伏安法(DPASV)；(d) 商品化 P25、TiO₂、g-C₃N₄ 和二维 TiO₂-g-C₃N₄ 样品的电化学阻抗谱(EIS)能斯特图

极和参比电极。首先,通过控制光的开/关周期(100 s)和外加电压(0.3 V vs. Ag/AgCl),在全波长照射下测定材料的光电流随时间的变化。如图9.10(b)所示,与 P25、TiO_2 和 $g-C_3N_4$ 样品的光电流相比,二维 $TiO_2-g-C_3N_4$ 材料的光电流明显增强。结果表明,异质结构的构建有利于促进光生载流子的分离和转移。光电流的形成主要是由于光诱导电子向背界面的扩散;同时,光诱导空穴与电解质中的受体反应。同时,DPV 被用于测定材料对 BPA 的电化学氧化性能。如图9.10(c)所示,与 P25、TiO_2 和 $g-C_3N_4$ 相比,2D $TiO_2-g-C_3N_4$ 复合材料具有更高的氧化电流密度和更低的氧化电位[41]。接着,电化学阻抗(EIS)(图9.10(d))显示所有样品都出现一个单一的电容电弧。并且,二维 $TiO_2-g-C_3N_4$ 中的电荷转移电阻明显小于 TiO_2、$g-C_3N_4$ 和商品化 P25 中的电荷转移电阻,表示电子转移电阻进一步降低,证明二维 $TiO_2-g-C_3N_4$ 异质结的协同效应有助于载流子的分离和运输。

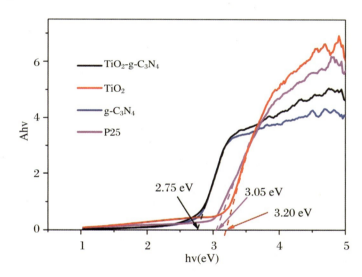

图9.11　根据所获得样品的 Ahv 与 hv 的作图曲线所计算的带隙值

9.5 二维二氧化钛与氮化碳复合材料光电催化双酚 A 降解性能及降解路径分析

为了评估所有样品的光电催化活性,所有制备的电极样品分别对 BPA 进行了光电催化降解实验。通过电化学工作站控制外加电压(与 SCE:0.8~1.3 V 相比),电解 50 mL 含 10 mg/L BPA 的 Na_2SO_4 水溶液(0.1 mol/L)。随后,在 PEC 反应期间,使用全波长的 300 W(12 A)氙灯(Microsolar 300,北京完美公司,中国)照亮每个 BPA 溶液。在带有冷浴的 60 mL 反应器中进行光电催化实验,并在给定时间间隔从反应器中滤出 1 mL 溶液,用于高效液相色谱测定。测定条件为:采用高效液相色谱法(HPLC-1100,美国安捷伦公司)测定双酚 A 的浓度。其中,该仪器配备了 Hypersil-ODS 反相柱和紫外检测器(波长 276 nm)。流动相为水和乙腈($V_{水}:V_{乙腈}$ = 45:55),流速为 0.8 mL/min。

基于此,根据不同外加电压下的实验(图 9.12(d))和先前的报道,选择 +1.3 V/SCE 作为合适的应用电压使用在光电催化系统中。空白对照实验组为无催化剂的碳纸电极(240 V 的紫外-可见光照射,施加电压为 +1.3 V)。图 9.12(a)显示其在 240 min 内的 BPA 降解效率(30%)。不同的是,在相同条件下,二维 TiO_2-g-C_3N_4/碳纸电极对 BPA 展现了快速的催化降解能力,在 240 min 内实现对 BPA 的最高降解效率(99.7%)(图 9.12(a))。然而,TiO_2、g-C_3N_4 和商用 P25 样品的总 BPA 光电催化降解率分别为 70.1%、80% 和 60%(图 9.12(a))。图 9.13 显示了使用二维 TiO_2-g-C_3N_4/碳纸电极对 BPA 污染物随时间变化的光电催化降解过程所测得的高效液相色谱图。进一步动力学分析,图 9.12(b)说明了随着时间变化的 BPA 光电催化降解过程,遵循伪一级动力学。因此,可以通过拟合曲线和计算 $\ln(C/C_0)$ 与时间(C 代表 BPA 浓度)的斜率来获得 BPA 降解速率的表观常数(k)。通过与 BPA 降解速率常数的比较,证实了用二维 TiO_2-g-C_3N_4/碳纸电极可以实现 BPA 的最快降解,具体为,$k_{TiO_2\text{-g-}C_3N_4}$ 的绝对值分别比 $k_{\text{g-}C_3N_4}$、k_{TiO_2} 和 k_{P25} 大 1.7 倍、2.5 倍和 3 倍。为了探索二维 TiO_2-g-C_3N_4 在光电催化系统中的光电结合优势,分别进行了光催化和电催化降解 BPA 实验(图 9.12(c))。结果表明,光电催化体系的 BPA 降解率比光催化和电催化两种体系对 BPA 降解率之和还高 2 倍左右。这表明,使用二维

TiO₂-g-C₃N₄/碳纸电极的光电催化系统可能存在光催化和电催化反应之间的协同效应，从而显著提高了 BPA 降解的效率，实现了光电催化"1+1＞2"的效果[10]。

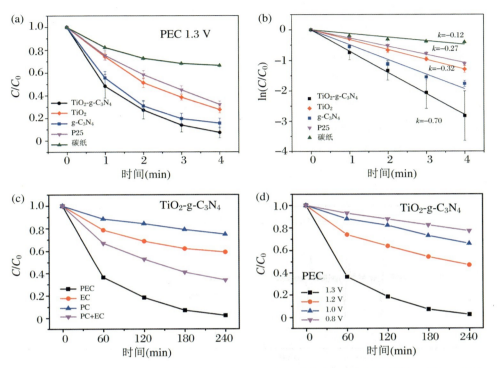

图 9.12　施加电压为 1.3 V 的全波长光照下 BPA 的 PEC 降解(a)和相关的伪一级动力学常数(b)；PC、EC 和 PEC 条件下 BPA 的去除(c)和使用二维 TiO₂-g-C₃N₄/碳纤维电极施加不同电压的 PEC(d)

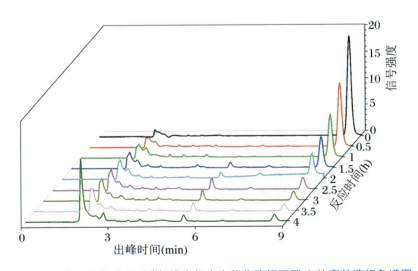

图 9.13　二维 TiO₂/g-C₃N₄/碳纤维电极光电催化降解双酚 A 的高效液相色谱图

在光电催化降解污染物过程中,有效的活性氧化物种被认为是降解矿化污染物的强氧化剂,需要进一步研究。二维 TiO_2-g-C_3N_4/碳纸电极将产生电子和空穴,这些电子和空穴迁移到催化剂的界面,与被吸收的 O_2 和 H_2O 反应,然后产生强活性氧化物质,如·OH 自由基。因此,ESR 光谱被用来测定该体系中是否存在·OH 自由基。在光电催化过程中,使用二维 TiO_2-g-C_3N_4/碳纸电极观察到明显的 4 个特征峰,强度比为 1∶2∶2∶1(图 9.14(a))。根据文献资料显示[42-43],证实这些峰属 DMPO-·OH 加合物。因此,·OH 自由基是该体系 BPA 污染物降解过程中的主要氧化活性物种。由于稳定的性能对实际废水处理的催化剂起着重要作用,二维 TiO_2-g-C_3N_4/碳纸电极被重复用于 BPA 降解,5 个周期后略有下降(图 9.14(b))。这表明,二维 TiO_2-g-C_3N_4 具有良好的稳定性,在实际应用中可作为一种优良的光电催化材料。

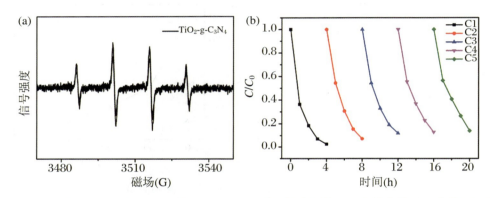

图 9.14　在 PEC 过程中,使用二维 TiO_2-g-C_3N_4/碳纤维电极的 DMPO 溶液的 ESR 光谱(a)以及 PEC 法降解 BPA 的二维 TiO_2-g-C_3N_4/碳纤维电极的循环稳定性实验(b)

9.6 二维二氧化钛与氮化碳复合材料的光电催化机制分析

基于以上实验结果,二维 TiO_2-g-C_3N_4 被鉴定为一种有效的光电催化剂。首先,二维 TiO_2-g-C_3N_4 表现出可见光活性,增强了在紫外-可见光照射下的光催化能力。其次，TiO_2-g-C_3N_4 异质结的二维结构有利于 g-C_3N_4 和 TiO_2 之间的粒子间电子传输，因为 TiO_2-g-C_3N_4 的荧光寿命比其他结构的荧光寿命慢。这表明具有较长荧光寿命的 TiO_2-g-C_3N_4 有利于更有效的电子转移(图 9.15(a))[25,27]。此外，根据 Ti—N 桥和 C—O 桥中的价带和导带位置，该方案还展示了电子转移途径(图 9.15(b))[25]。最后，根据 EIS 和 DFT 的结果，确认了 TiO_2-g-C_3N_4 异质结的二维结构可以进一步降低电子转移电阻，这有利于增加光催化和电催化的协同效应，提高光电催化性能。

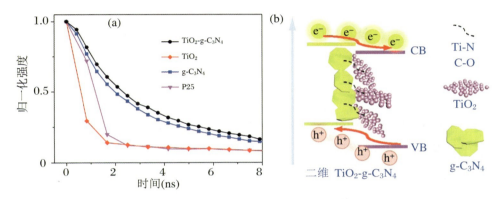

图 9.15　(a) 具有二维 TiO_2-g-C_3N_4、TiO_2、g-C_3N_4 和 P25 的发光寿命的发光衰减曲线；(b) Ti-N 和 C-O 桥接的二维 TiO_2-g-C_3N_4 中光生电荷转移和分离的示意图

我们成功地构建了 Ti—N 和 C—O 键连接的二维 TiO_2-g-C_3N_4，并将其应用于光电催化协同降解污染物。二维 TiO_2-g-C_3N_4/碳纸电极显示出较高的光电催化降解 BPA 的活性，其反应速率是光电催化系统中商品化 P25 的 3 倍。光电催化降解 BPA 性能是光催化和电催化性能总和的 2 倍左右。二维 TiO_2-g-C_3N_4 突出的光电催化性能优势主要归因于有效的电子-空穴分离和转移以及异质结的良好导电性。通过荧光寿命的测定，不仅证实了 TiO_2 与 g-C_3N_4 之间的电荷分离和输运增强；而且根据 EIS 和 DFT 计算结果表明，得到的 TiO_2-g-C_3N_4 异质结的导电性也明显提高。因此，通过提高光催化和电催化活性，增强了 TiO_2-g-C_3N_4 异质结光电催化性能的协同效应。这项工作为设计二维结构的高效光电催化材料提供了新的见解，通过同时提高了其光物理和光电化学性能，达到高效的光电催化性能，并有望在实际污水处理中得到应用。

参考文献

[1]　LINSEBIGLER A L, LU G, YATES J T. Photocatalysis on TiO_2 surfaces:

principles, mechanisms, and selected results[J]. Chem. Rev., 1995, 95: 735-758.

[2] SABZEHPARVAR M, KIANI F, TABRIZI N S. Mesoporous-assembled TiO_2-NiO-Ag nanocomposites with p-n/schottky heterojunctions for enhanced photocatalytic performance[J]. J. Alloys Compd., 2021, 876. DOI: 10.101b/j.jaUcom.2021.160133.

[3] LIU G, YANG H G, PAN J, et al. Titanium dioxide crystals with tailored facets[J]. Chem. Rev., 2014, 114:9559-9612.

[4] YANG H G, SUN C H, QIAO S Z, et al. Anatase TiO_2 single crystals with a large percentage of reactive facets[J]. Nature, 2008, 453:638-641.

[5] WANG W-K, CHEN J-J, LI W-W, et al. Synthesis of Pt-loaded self-interspersed anatase TiO_2 with a large fraction of {001} facets for efficient photocatalytic nitrobenzene degradation[J]. ACS Appl. Mater. Interfaces, 2015, 7:20349-20359.

[6] WANG W-K, CHEN J-J, GAO M, et al. Photocatalytic degradation of atrazine by boron-doped TiO_2 with a tunable rutile/anatase ratio[J]. Appl. Catal. B-Environ., 2016, 195:69-76.

[7] WANG J, YANG J K, ZHAO Y M, et al. Photoelectrochemical properties of porous Si/TiO_2 nanowire heterojunction structure[J]. J. Nanomater., 2020, 2020:1-7.

[8] LAL S D, JOSE T S, RAJESH C, et al. Accelerated photodegradation of polystyrene by TiO_2-polyaniline photocatalyst under UV radiation[J]. Eur. Polym. J., 2021, 153. DOI: 10.1016/j.eurpolymj.2021.11493.

[9] LIU C, ZHANG A-Y, SI Y, et al. Photochemical anti-fouling approach for electrochemical pollutant degradation on facet-tailored TiO_2 single crystals [J]. Environ. Sci. Technol., 2017, 51:11326-11335.

[10] LIU C, ZHANG A-Y, PEI D-N, et al. Efficient rlectrochemical reduction of nitrobenzene by defect-engineered TiO_{2-x} single crystals[J]. Environ. Sci. Technol., 2016, 50:5234-5242.

[11] WANG G X, ZHANG H L, WANG W Z, et al. Fabrication of Fe-TiO_2-NTs/SnO_2-Sb-Ce electrode for electrochemical degradation of aniline[J]. Sep. Purif. Technol., 2021, 268.

[12] KOO M S, CHO K, YOON J, et al. Photoelectrochemical degradation of organic compounds coupled with molecular hydrogen generation using elec-

trochromic TiO$_2$ nanotube arrays[J]. Environ. Sci. Technol., 2017, 51: 6590-6598.

[13] PENG Y-P, CHEN H, HUANG C P. The synergistic effect of photoelectrochemical (PEC) reactions exemplified by concurrent perfluorooctanoic acid (PFOA) degradation and hydrogen generation over carbon and nitrogen codoped TiO$_2$ nanotube arrays (C-N-TNTAs) photoelectrode[J]. Appl. Catal. B-Environ., 2017, 209:437-446.

[14] XU Y, HE Y, CAO X, et al. TiO$_2$/Ti rotating disk photoelectrocatalytic (PEC) reactor: a combination of highly effective thin-film PEC and conventional PEC processes on a single electrode[J]. Environ. Sci. Technol., 2008, 42:2612-2617.

[15] XU Y, JIA J, ZHONG D, et al. Degradation of dye wastewater in a thin-film photoelectrocatalytic (PEC) reactor with slant-placed TiO$_2$/Ti anode [J]. Chem. Eng. J., 2009, 150:302-307.

[16] YAO Y, LI K, CHEN S, et al. Decolorization of Rhodamine B in a thin-film photoelectrocatalytic (PEC) reactor with slant-placed TiO$_2$ nanotubes electrode[J]. Chem. Eng. J., 2012, 187:29-35.

[17] CHEN W, WANG T, XUE J, et al. Cobalt-nickel layered double hydroxides modified on TiO$_2$ nanotube arrays for highly efficient and stable PEC water splitting[J]. Small, 2017, 13:1602420.

[18] WILKE C M, WUNDERLICH B, GAILLARD J-F O, et al. Synergistic bacterial stress results from exposure to nano-Ag and nano-TiO$_2$ mixtures under light in environmental media[J]. Environ. Sci. Technol., 2018, 52: 3185-3194.

[19] LIU Y, XIE C, LI H, et al. Low bias photoelectrocatalytic (PEC) performance for organic vapour degradation using TiO$_2$/WO$_3$ nanocomposite [J]. Appl. Catal. B-Environ., 2011, 102:157-162.

[20] ZHANG J, CHEN Y, WANG X. Two-dimensional covalent carbon nitride nanosheets: synthesis, functionalization, and applications[J]. Energ. Environ. Sci., 2015, 8:3092-3108.

[21] LIU W Z, SUN M X, DING Z P, et al. Ti$_3$C$_2$ MXene embellished g-C$_3$N$_4$ nanosheets for improving photocatalytic redox capacity[J]. J. Alloys Compd., 2021, 877. DOI: 10.1016/j.jallcom.2021.160223.

[22] GUO S, ZHAO S, WU X, et al. A Co$_3$O$_4$-CDots-C$_3$N$_4$ three component

electrocatalyst design concept for efficient and tunable CO_2 reduction to syngas[J]. Nat. Commun., 2017, 8:1828-1828.

[23] BIAN H, WANG A, LI Z, et al. g-C3N4-modified water-crystallized mesoporous SnO_2 for enhanced photoelectrochemical properties[J]. Part. Part. Syst. Char., 2018, 35. DOI: 10.1002/PPSC.202800155.

[24] WANG X, WANG G, CHEN S, et al. Integration of membrane filtration and photoelectrocatalysis on g-C_3N_4/CNTs/Al_2O_3 membrane with visible-light response for enhanced water treatment[J]. J. Membrane. Sci., 2017, 541:153-161.

[25] SHI X, FUJITSUKA M, LOU Z, et al. In situ nitrogen-doped hollow-TiO_2/g-C_3N_4 composite photocatalysts with efficient charge separation boosting water reduction under visible light[J]. J. Mater. Chem. A, 2017, 5: 9671-9681.

[26] SU J, GENG P, LI X, et al. Novel phosphorus doped carbon nitride modified TiO_2 nanotube arrays with improved photoelectrochemical performance [J]. Nanoscale, 2015, 7:16282-16289.

[27] ELBANNA O, FUJITSUKA M, MAJIMA T. g-C_3N_4/TiO_2 mesocrystals composite for H_2 evolution under visible-light irradiation and its charge carrier dynamics[J]. ACS Appl. Mater. Interfaces., 2017, 9:34844-34854.

[28] LIU C, RAZIQ F, LI Z, et al. Synthesis of TiO_2/g-C_3N_4 nanocomposites with phosphate-oxygen functional bridges for improved photocatalytic activity [J]. Chinese J. Catal., 2017, 38:1072-1078.

[29] LI C, SUN Z, ZHANG W, et al. Highly efficient g-C_3N_4/TiO_2/kaolinite composite with novel three-dimensional structure and enhanced visible light responding ability towards ciprofloxacin and S. aureus[J]. Appl. Catal. B-Environ., 2018, 220:272-282.

[30] JO W-K, NATARAJAN T S. Influence of TiO_2 morphology on the photocatalytic efficiency of direct Z-scheme g-C_3N_4/TiO_2 photocatalysts for isoniazid degradation[J]. Chem. Eng. J., 2015, 281:549-565.

[31] LI G, LIAN Z, WANG W, et al. Nanotube-confinement induced size-controllable g-C_3N_4 quantum dots modified single-crystalline TiO_2 nanotube arrays for stable synergetic photoelectrocatalysis[J]. Nano Energy, 2016, 19:446-454.

[32] CAI X, MAO L, YANG S, et al. Ultrafast charge separation for full solar

spectrum-activated photocatalytic H_2 generation in a black phosphorus-Au-CdS heterostructure[J]. ACS Energy Lett., 2018, 3:932-939.

[33] WANG X-J, TIAN X, SUN Y-J, et al. Enhanced schottky effect of a 2D-2D CoP/g-C_3N_4 interface for boosting photocatalytic H_2 evolution[J]. Nanoscale, 2018, 10:12315-12321.

[34] LIU J. Origin of high photocatalytic efficiency in monolayer g-C_3N_4/CdS heterostructure: a hybrid DFT study[J]. J. Phys. Chem. C, 2015, 119: 28417-28423.

[35] MAO L, CAI X, YANG S, et al. Black phosphorus-CdS-$La_2Ti_2O_7$ ternary composite: effective noble metal-free photocatalyst for full solar spectrum activated H_2 production[J]. Appl. Catal. B-Environ., 2019, 242:441-448.

[36] ERDEM B, HUNSICKER R A, SIMMONS G W, et al. XPS and FTIR surface characterization of TiO_2 particles used in polymer encapsulation[J]. Langmuir, 2001, 17:2664-2669.

[37] SONG X, LI W, HE D, et al. The "Midas Touch" transformation of TiO_2 nanowire arrays during visible light photoelectrochemical performance by carbon/nitrogen coimplantation[J]. Adv. Energy Mater., 2018, 8:1800165.

[38] KONG L, ZHANG X, WANG C, et al. Ti^{3+} defect mediated g-C_3N_4/TiO_2 Z-scheme system for enhanced photocatalytic redox performance[J]. Appl. Surf. Sci., 2018, 448:288-296.

[39] LI Y, JIN R, XING Y, et al. Macroscopic foam-like holey ultrathin g-C_3N_4 nanosheets for drastic improvement of visible-light photocatalytic activity [J]. Adv. Energy Mater., 2016, 6: 1601273.

[40] ABDELRAHEEM W H M, PATIL M K, NADAGOUDA M N, et al. Hydrothermal synthesis of photoactive nitrogen and boron-codoped TiO_2 nanoparticles for the treatment of bisphenol A in wastewater: synthesis, photocatalytic activity, degradation byproducts and reaction pathways[J]. Appl. Catal. B-Environ., 2019, 241:598-611.

[41] PEI D-N, ZHANG A-Y, PAN X-Q, et al. Electrochemical sensing of bisphenol a on facet-tailored TiO_2 single crystals engineered by inorganic-framework molecular imprinting sites[J]. Anal. Chem., 2018, 90: 3165-3173.

[42] HARBOUR J R, CHOW V, BOLTON J R. An electron spin resonance study

of the spin adducts of OH and HO_2 radicals with nitrones in the ultraviolet photolysis of aqueous hydrogen peroxide solutions[J]. Can. J. Chem., 1974, 52: 3549-3553.

[43] SARGENT F P, GARDY E M. Spin trapping of radicals formed during radiolysis of aqueous solutions. Direct electron spin resonance observations [J]. Can. J. Chem., 1976, 54: 275-279.

第 10 章

单颗粒荧光光谱
和单分子成像技术解析
二氧化钛光催化高活性位点

10.1
二氧化钛催化机制的原位光谱研究进展

TiO$_2$作为一种完美的光催化剂从合成方法和催化性能方面已经被大量研究[1-4]。进而,大量的显微镜被用于研究 TiO$_2$ 光催化反应及其相关的具体反应机制,并发展相关催化剂[5-9]。在已有的报道中,荧光显微镜被用于从单颗粒单分子水平研究 TiO$_2$ 光催化反应,并且该实验证明了在单个锐钛矿 TiO$_2$ 上的光催化还原位点和晶面的关系[10]。同时,扫描隧道显微镜则被用于原位研究 TiO$_2$ 表面氧的吸附行为[11]、CO 吸附位点[12]、光催化水的分解反应[13]和缺陷的作用等[14-15]。最近,TiO$_2$ 纳米棒上负载助催化剂研究电子和空穴发生的氧化还原空间相关性则使用超分辨空穴和电子反应图[16]。另外,单颗粒荧光光谱测试被认为是研究催化剂电子空穴对的形成、传递、分离和复合[10,17-22]最直接的技术。因此,单分子和单颗粒对活性位点的检测用于研究 TiO$_2$ 上光生载流子的分离和转移。

观测的荧光寿命不仅依赖于{101}晶面和{001}晶面间的带隙[23-26],而且也受晶面缺陷影响[14,27]。所以荧光寿命和材料结构活性的相关性需要更直接的研究。因此,单个锐钛矿 TiO$_2$ 上的载流子的移动和复合,可以通过空间和时间分辨的荧光测量进行研究。

在本章的研究中,我们使用单颗粒单分子荧光显微镜(200 nm 的空间分辨率和 0.1~0.2 ns 的时间分辨)对锐钛矿 TiO$_2$ 纳米颗粒的荧光强度及寿命进行表征研究。我们发现 TiO$_2$ 对于染料分子反应的活性位点,非常吻合单颗粒测试过程中发现单颗粒空间分辨率的条件下,载流子的复合位点和荧光寿命。从而,实验证明了对于单个 TiO$_2$ 纳米颗粒,光生载流子更趋向于晶面间的区域(TiO$_2$ 边或角)进行反应,作为高反应活性位点。

10.2 高暴露晶面锐钛矿二氧化钛的合成与表征

10.2.1 高暴露晶面锐钛矿二氧化钛的制备

本章工作中所用化学试剂均为分析纯级别，且未做进一步纯化。锐钛矿 TiO_2 纳米颗粒是根据先前报道的方法合成[28]。首先，65 mg 的 $TiOSO_4$ 溶解在装有 40 mL 氢氟酸（120 mmol/L）水溶液烧杯中。等搅拌 30 min 后，将溶液转移至高压釜中。接着，在电烘箱中，进行 180 ℃ 水热反应 12 h。在材料生长完成后，高压釜冷却至室温。所获得的样品依次在乙醇和水溶液中超声清洗。最后，在单分子和单颗粒测试前，获得的锐钛矿 TiO_2 纳米颗粒需要进行 0.1 mol/L NaOH 溶液清洗去除吸附的 F^-。

10.2.2 单分子和单颗粒荧光测试实验的样品准备

石英盖玻片（DAICO MFG CO 公司，日本）经过洗涤剂（As One，Cleanace）超声清洗 7 h，接着用清水连续清洗 5 次。最后石英盖玻片通过超纯水（millipore）进行清洗。

为了获得单分散的锐钛矿 TiO_2 纳米颗粒，将甲醇水溶液中的具有良好分散性低浓度 TiO_2 旋涂到干净的盖玻片上。接着，将该颗粒负载的盖玻片在烘箱中，进行 1 h 的固定（100 ℃）。

10.2.3
高暴露晶面锐钛矿二氧化钛的物相及微观结构分析

首先,使用衍射仪(Rigaku 公司,日本)测试得到 XRD 图谱,使用 Cu Kα 放射线(λ = 1.541841 × 10^{-10} m)在 8°/min 条件下测试。图 10.1(a)所示的合成样品的 XRD 图,非常吻合锐钛矿标准卡片,表明该合成样品为锐钛矿,并且无其他物相峰出现[29]。同时,TiO_2 纳米结构的形貌分别通过场发射扫描电子显微镜(JEOL 公司,日本)和透射电子显微镜(JEOL 公司,日本,在加速电压 200 kV)得

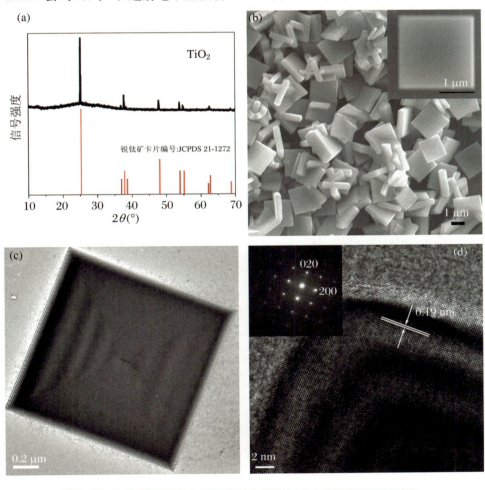

图 10.1 锐钛矿 TiO_2 的 XRD 图(a),SEM 图(b),TEM 图(c),高分辨 TEM 图(d)
图(b)的内插图为单个颗粒,图(d)的内插图为选区电子衍射图

到高分辨扫描电镜图和透射电子图像。图 10.1(b)显示该锐钛矿 TiO_2 颗粒均匀,平均尺寸为 2~3 μm。图 10.1(b)的内插图和图 10.1(c)的透射电镜图显示该 TiO_2 颗粒表面平滑,无小颗粒附着。而高分辨显微镜图(图 10.1(d))的一维晶格条纹证明该 TiO_2 结晶性良好,并且通过软件(Digital Micrograph)测量出晶格条纹间距为 0.19 nm,与锐钛矿 TiO_2 的{001}晶面相对应[24]。图 10.1(d)内插图的选取电子衍射确认了该锐钛矿 TiO_2 高暴露{001}晶面[30]。

10.2.4
选择性氧化和还原的荧光探针

荧光探针反应包含中性红(Amplex red)的氧化和刃天青(Rezazurin)的还原,并且它们反应后都可以形成荧光产物(Resorufin),如图 10.2(a、d)所示[16]。还原反应探测光生电子,氧化反应探测光生空穴。图 10.2(b、c)显示光催化还原反应过程的紫外-可见吸收和荧光光谱。从紫外-可见吸收光谱可以看出,在光催化还原刃天青的过程中,570 nm 处峰强度增加,相应 604 nm 处原始的刃天青峰强度下降。同样,在荧光光谱中 570 nm 处峰的强度随着反应过程增加。同理,图 10.2(e、f)显示光催化氧化反应过程的紫外-可见吸收和荧光光谱。从紫外-可见吸收光谱可以看出,在光催化氧化中性红的过程中,570 nm 处峰强度增加。并且 570 nm 处的荧光峰同样增加。

上述的荧光反应基于光催化实验和先前的文献报道[16]。所以,本实验可以通过对该反应产物的荧光检测,从单分子水平来研究单个颗粒上的界面电子传递。

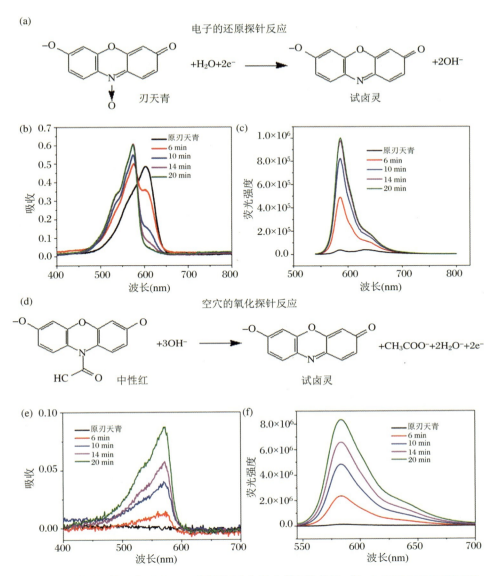

图 10.2 荧光探针反应:(a) 刃天青的还原;(d) 中性红的氧化;(b、e) 刃天青和中性红发生反应的后的紫外-可见吸收图;(c、f) 荧光图

10.3 在宽场显微镜下单分子反应的荧光测试

实验装置基于奥林巴斯 IX71 倒置荧光显微镜。通过使用卤素灯（Olympus，U-LH100L-3）照射样品获得的透射图像来确定固定在盖玻片上的锐钛矿 TiO_2 颗粒的位置。从 CW Ar 离子激光器（Melles Griot，IMA101010BOS；532 nm，玻璃表面能量密度的 $0.1\ kW/cm^{-2}$）发射的圆偏振光利用第一分色镜（Olympus，DM505）朝向第二分色镜（Olympus，DM505）反射，RDM450），其反射波长长于 450 nm，对于小于 450 nm 的波长是透明的。通过第二分色镜反射后通过物镜（Olympus，UPLSAPO 100XO；1.40NA，100×）的激光在盖玻璃-甲醇界面处完全反射。这导致了消逝场的产生，从而可以检测到单分子荧光染料信号。对于 TiO_2 晶体的激发，使用来自 CW Ar 离子激光器（Melles Griot，IMA101010BOS；532 nm，玻璃表面为 $0.1\ kW/cm^2$）的 405 nm 光。使用相同的目的收集在玻璃上的锐钛矿 TiO_2 颗粒上产生的荧光产物的发射，然后通过 1.6 倍内置放大变换器放大，通过带通滤光器（Semrock，FF01-575/30）去除不期望的散射光，然后使用电子倍增电荷耦合器件（EM-CCD）相机（Roper Scientific，Cascade Ⅱ:512）进行成像。以 20 帧/s 的帧速记录图像，并使用 ImageJ（http://rsb.info.nih.gov/ij/）或 OriginPro 8.5（OriginLab）进行处理。所有实验数据均在室温下得到。

全内反射荧光（TIRF）显微镜被用于单分子水平上研究在单个 TiO_2 颗粒晶面相关的光催化还原刃天青和氧化中性红实验[10]。图 10.3、图 10.4 分别显示锐钛矿 TiO_2 纳米颗粒在 532 nm 和 405 nm 双激光照射下，在水甲醇或水溶液中，染料分子发生反应的典型 TIRF 图。图 10.3(a)、图 10.4(a)显示在光照下没有染料分子下的 TiO_2 纳米颗粒 TIRF 图，作为背景对照。当有染料分子时，大量的亮点被观察到。通过质心分析确定发生反应的活性位点。有趣的是，在氧化和还原实验过程中，发现荧光亮点更倾向与发生在晶面界面（棱和角），如图 10.3(b～f)、图 10.4(b～f)所示。这个结果显示晶面对光催化活性有着重大的影响，但不涉及材料内部的活性评估。值得注意的是，在晶面界面（棱和角）更可能作为高活性位点。

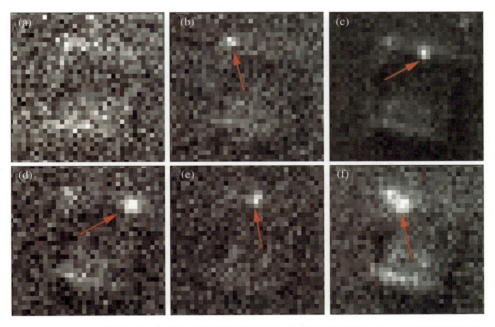

图 10.3 (a) 在单晶 TiO_2 上的光催化还原过程中的单分子图片背景;(b~f) 反应的不同时间点

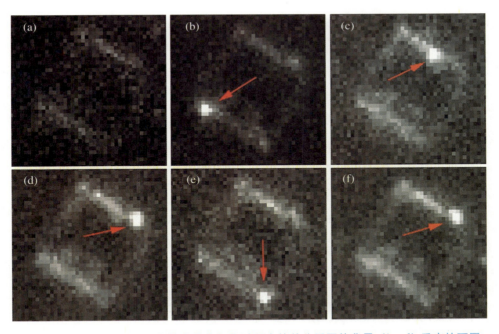

图 10.4 (a) 在单晶 TiO_2 上的光催化氧化过程中的单分子图片背景;(b~f) 反应的不同时间点

10.4 共聚焦显微镜下单颗粒荧光成像测定

使用与 Olympus IX71 倒置荧光显微镜联用的共聚焦显微镜系统（MicroTime 200，PicoQuant 公司，德国）扫描记录单粒子样品的 PL 图像和光谱。通过浸油物镜（UplanSApochromat，100×，1.4NA Olympus 公司，日本）激发样品，并使用再生扩增的钛蓝宝石激光驱动器（Spectra-Physics，Spitfire Pro F，1 kHz，Newport 公司，美国）通过 Nd:YLF 激光器（Spectra-Physics，Empower 15，Newport 公司，美国）泵浦。样品的 PL 测量的典型激励功率为 350 μW。通过相同的目标收集样品的发射，并通过二光分束器（Chroma，405rdc）和长通滤光器（Chroma，HQ430CP），最后由单光子雪崩光电二极管（Micro Photon Devices，PDM 50CT）进行检测。对于光谱学，仅通过配备有成像光谱仪（Acton Research，SP-2356）的 EM-CCD 照相机（Princeton Instruments，ProEM）检测通过狭缝的发射。光谱通常整合 30^{-1} s。使用个人计算机进行存储和分析由 EMCCD 摄像机检测到的频谱。所有实验数据均在室温下得到。

在单分子研究的基础上，需要分析单个锐钛矿 TiO_2 纳米颗粒具体光催化反应机制[5-6]。这里，单颗粒显微镜用于研究单个锐钛矿 TiO_2 纳米颗粒的电子传递和荧光寿命。图 10.5(a)显示的是典型单个锐钛矿 TiO_2 纳米颗粒的荧光图。从单颗粒荧光图可以看出，在锐钛矿 TiO_2 纳米颗粒中间的荧光强度很弱。相反，在{001}和{101}晶面结表面周围的荧光强度明显高于其他地方。这个现象表明，更多的电子-空穴对在异质结表面激发和复合[23]。

此外，图 10.5(c)显示锐钛矿 TiO_2 纳米颗粒上不同点的具体荧光光谱。根据文献报道，在 630 nm 处的发射峰是源于锐钛矿 TiO_2 纳米颗粒表面的载流子复合。从图中可以看出，荧光强度从中心（S_m）到边（E_{101}）逐渐增加。这些结果表明锐钛矿 TiO_2 晶面界面处的光催化活性最高。

为了进一步研究锐钛矿 TiO_2 界面的电子传递，图 10.5(d)显示了该颗粒的表面各处的荧光衰减曲线。结果清晰地显示在单个锐钛矿 TiO_2 颗粒上载流子寿命的空间变化。从点 S_m 到点 E_{101}，荧光寿命逐渐增加，符合荧光光谱的结果。而且 E_{001} 和 E_{101} 两点处的荧光寿命相近，表明在晶面界面处为高活性位点。这是因为在该区域存在表面异质结，可以促进光生载流子的分离[23,31]。表 10.1 则具体显示单个锐钛矿 TiO_2 颗粒上各个点的荧光寿命。

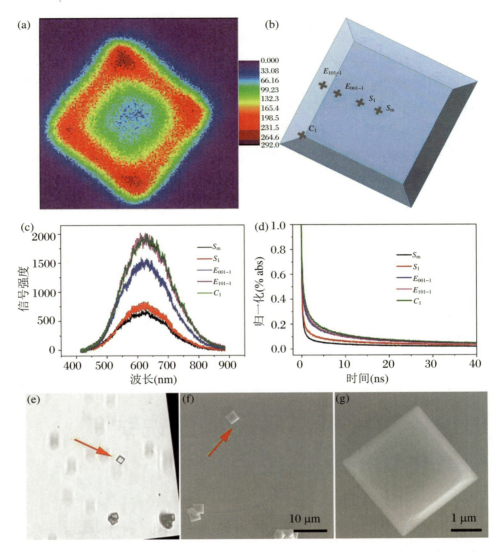

图 10.5　单个锐钛矿 TiO_2 颗粒的荧光图(a)、示意图(b)、荧光光谱(c)、电子寿命(d)、相应的显微镜图(e)、SEM 图(f)以及高分辨 SEM 图(g)

表 10.1　锐钛矿 TiO_2 颗粒上各个点的荧光寿命

	τ_1(ps)	比例(%)	τ_2(ps)	比例(%)	τ_3(ps)	比例(%)
S_m	0.14	82	1.71	13	23.22	5
S_{001-1}	0.11	74	1.59	18	22.79	7
E_{001-1}	0.14	64	2.09	24	23.92	12
E_{101-1}	0.18	62	2.23	26	23.31	12
C_1	0.18	56	2.23	31	22.83	14

图 10.5(e)显示相对应荧光图单个锐钛矿 TiO_2 颗粒的典型显微镜图(红色箭头所示)。除此之外,单独在石英片上的 TiO_2 颗粒通过 SEM 图进一步确认,如图 10.5(f)和图 10.5(g)所示。

根据锐钛矿 TiO_2 颗粒活性位的位置,各个点(S_m 到 C_1)的平均寿命分别为 1.50 ns、1.86 ns、3.05 ns、3.45 ns 和 4.15 ns。这个结果符合单颗粒锐钛矿 TiO_2 的荧光强度。这也进一步证实了我们先前的假设:主要的电子-空穴对形成传递在锐钛矿 TiO_2 表面异质结附近(边和棱)。这些结果证明更多的活性位点存在在锐钛矿 TiO_2 表面异质结区域,与先前单分子氧化还原实验的结果相一致。

基于以上的分析,光生电子-空穴对在单个锐钛矿 TiO_2 上的传递机制变得清晰。正是由于存在导带位置的能量差,更多的光生电子可以从{001}到{101}晶面的晶面异质结传递,然后与光生空穴复合[23]。这样,电子的荧光寿命更强,区域的荧光强度也增强。

10.5 理论计算模拟高暴露晶面锐钛矿二氧化钛的活性位点

为了更好地解析单个锐钛矿型 TiO_2 颗粒中{001}表面、{101}表面和晶体边缘的光催化活性,并进一步了解不同界面的性质,荧光探针反应过程在模型上通过 DFT 计算模拟。团簇模型的状态密度(DOS)表明,晶体边缘显示出不同于{001}表面和{101}表面的能带结构的其他特征。带隙的绝对值被密度泛函方法低估了,但这些带的相对能级仍可用于理论分析。如图 10.6(b)和(c)所示,晶体边缘的导带最小值(CBM)和价带最大值(VBM)低于{001}表面的。因此,光生电子(e^-)能够从{001}表面流动到边缘部位,而空穴(h^+)向相反的方向移动,促进了分离-光激发的 h^+/e^- 的比率。在该连接处,边缘位点用作刃天青还原的光催化还原位点。同时,{101}表面的 CBM 和 VBM 均低于晶体边缘的 CBM 和 VBM,电子结构的结果表明,由{001}表面和{101}表面形成的晶体边缘积累了光生电子和空穴,并充当氧化还原反应位点。

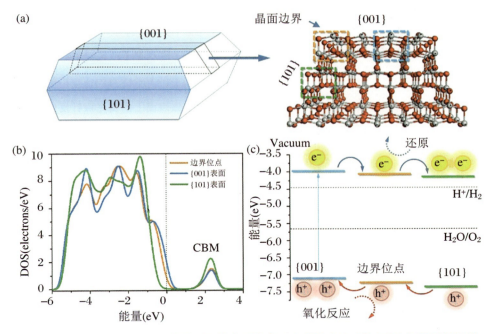

图 10.6 锐钛矿型 TiO_2 的团簇模型及其电子结构:(a) 具有{001}和{101}表面及其晶体边缘的锐钛矿型 TiO_2 模型(虚线框中的位点用于分析 DOS);(b) 边缘位置,群集模型的{001}曲面和{101}曲面的 DOS;(c) 从 DOS 中 VBM 和 CBM 的位置以及流动到不同光催化部位的能带的相对能级光生载流子的路径

此外,计算了锐钛矿型 TiO_2 的表面和晶体边缘的荧光探针反应过程 Gibbs 自由能变化(ΔG)和能垒(E_a)如表 10.2 所示。先前的工作表明,大部分锐钛矿型 TiO_2 中的光生电子和空穴分别迁移到{101}面和{001}面,然后参与光催化反应的还原和氧化过程。因此,为了进一步发现还原位点,在晶体边缘和{101}晶面之间比较了刃天青的还原能力。负 ΔG 值越负证明晶体边缘具有更高的增强能力常数,这归因于晶体边缘处更多的光生电子的积累。此外,类似的分析表明,由于光生空穴多于{001}表面,晶体边缘也促进了双链红色氧化反应以达到平衡状态。同时,荧光探针反应的较低 E_a,进一步表明晶体边缘是具有较高反应速率的光催化反应的动力学可行位点。因此,在原子水平上的理论计算表明,晶体边缘的独特几何构型提高了光生载流子用于光催化的俘获概率,与我们的实验结果一致。

在本章中,我们实现了锐钛矿 TiO_2 颗粒的空间和时间分辨的荧光测量(电荷载流子迁移和复合的差异),并在单颗粒上表征了锐钛矿 TiO_2 颗粒的光致发光强度和寿命。此外,原位单分子荧光检测证实了光催化氧化还原活性位点基本位于锐钛矿 TiO_2 颗粒的晶面附近,同时单颗粒的原位表征手段显示位于

{101}和{001}晶面之间荧光强度更强和寿命更长。由此说明锐钛矿 TiO_2 颗粒的光催化活性位于{101}和{001}晶面附近，对光催化活性起主导作用。这种单分子单颗粒方法既可以对 TiO_2 光催化反应机制进行微观解析，又适用于在非均相催化中及在生物系统的实时观测。

表 10.2　计算锐钛矿型 TiO_2{001}、{101}及其晶体边缘表面上的荧光探针反应热力学(Gibbs 自由能变化 ΔG)和动力学特点(能垒 E_a)

反应	反应位点	ΔG (eV)	E_a (eV)
反应式 1（+H_2O+2e⁻ → +2OH⁻）	{101} surface	−0.263	4.66
	Crystal edge	−0.657	2.82
反应式 2（+H_2O+2e⁻ → +2OH⁻）	{001} surface	−1.10	3.54
	Crystal edge	−1.60	−0.22*

* 负活化能代表通过捕获势阱中的分子来进行无能垒、高反应速率反应。

参考文献

[1] HOFFMANN M R, MARTIN S T, CHOI W, et al. Environmental applications of semiconductor photocatalysis[J]. Chem. Rev., 1995, 95:69-96.

[2] FUJISHIMA A, HONDA K. Electrochemical photolysis of water at a semiconductor electrode[J]. Nature., 1972, 238:37-38.

[3] LIU G, YANG H G, PAN J, et al. Titanium dioxide crystals with tailored facets[J]. Chem. Rev., 2014, 114:9559-9612.

[4] XING Z, ZHANG J, CUI J, et al. Recent advances in floating TiO_2-based photocatalysts for environmental application[J]. Appl. Catal. B., 2018, 225:452-467.

[5] HENDERSON M A, LYUBINETSKY I. Molecular-level insights into photocatalysis from scanning probe microscopy studies on TiO_2{110}[J]. Chem. Rev., 2013, 113:4428-4455.

[6] LINSEBIGLER A L, LU G, YATES J T, JR. Photocatalysis on TiO_2

surfaces: principles, mechanisms, and selected results[J]. Chem Rev., 1995, 95:735-758.

[7] SCHNEIDER J, MATSUOKA M, TAKEUCHI M, et al. Understanding TiO_2 photocatalysis: mechanisms and materials[J]. Chem. Rev., 2014, 114:9919-9986.

[8] LIU X, DONG G, LI S, et al. Direct observation of charge separation on anatase TiO_2 crystals with selectively etched {001} facets[J]. J. Am. Chem. Soc., 2016, 138:2917-2920.

[9] FANG K, LI G, OU Y, et al. An environmental transmission electron microscopy study of the stability of the TiO_2 (1×4) reconstructed {001} surface[J]. J. Phys. Chem. C, 2019, 123:21522-21527.

[10] TACHIKAWA T, YAMASHITA S, MAJIMA T. Evidence for crystal-face-dependent TiO_2 photocatalysis from single-molecule imaging and kinetic analysis[J]. J. Am. Chem. Soc., 2011, 133:7197-7204.

[11] TAN S, JI Y, ZHAO Y, et al. Molecular oxygen adsorption behaviors on the rutile TiO_2{110}-1×1 surface: An in situ study with low-temperature scanning tunneling Microscopy[J]. J. Am. Chem. Soc., 2011, 133:2002-2009.

[12] ZHAO Y, WANG Z, CUI X, et al. What are the adsorption sites for CO on the reduced TiO_2{110}-1×1 surface?[J]. J. Am. Chem. Soc., 2009, 131:7958-7959.

[13] TAN S, FENG H, JI Y, et al. Observation of photocatalytic dissociation of water on terminal Ti sites of TiO_2{110}-1×1 surface[J]. J. Am. Chem. Soc., 2012, 134:9978-9985.

[14] WANG Y, SUN H, TAN S, et al. Role of point defects on the reactivity of reconstructed anatase titanium dioxide {001} surface[J]. Nat. Commun., 2013, 4:2214.

[15] LI G, FANG K, OU Y, et al. Surface study of the reconstructed anatase TiO_2{001} surface[J]. Prog. Nat. Sci-Mater., 2021, 31:1-13.

[16] SAMBUR J B, CHEN T Y, CHOUDHARY E, et al. Sub-particle reaction and photocurrent mapping to optimize catalyst-modified photoanodes[J]. Nature, 2016, 530:77-80.

[17] LOU Z, FUJITSUKA M, MAJIMA T. Two-dimensional Au-nanoprism/reduced graphene oxide/Pt-nanoframe as plasmonic photocatalysts with

multiplasmon modes boosting hot electron transfer for hydrogen generation [J]. J. Phys. Chem. Lett., 2017, 8:844-849.

[18] ZHU M, CAI X, FUJITSUKA M, et al. Au/$La_2Ti_2O_7$ nanostructures sensitized with black phosphorus for plasmon-enhanced photocatalytic hydrogen production in visible and near-infrared light[J]. Angew. Chem. Int. Ed., 2017, 56:2064-2068.

[19] ZHENG Z, TACHIKAWA T, MAJIMA T. Single-particle study of Pt-modified Au nanorods for plasmon-enhanced hydrogen generation in visible to near-infrared region[J]. J. Am. Chem. Soc., 2014, 136:6870-6873.

[20] ZHENG Z, TACHIKAWA T, MAJIMA T. Plasmon-enhanced formic acid dehydrogenation using anisotropic Pd-Au nanorods studied at the single-particle level[J]. J. Am. Chem. Soc., 2015, 137:948-957.

[21] LOU Z, FUJITSUKA M, MAJIMA T. Pt-Au triangular nanoprisms with strong dipole plasmon resonance for hydrogen generation studied by single-particle spectroscopy[J]. ACS Nano, 2016, 10:6299-6305.

[22] LOU Z, KIM S, ZHANG P, et al. In situ observation of single Au triangular nanoprism etching to various shapes for plasmonic photocatalytic hydrogen generation[J]. ACS Nano, 2017, 11:968-974.

[23] YU J, LOW J, XIAO W, et al. Enhanced photocatalytic CO_2-reduction activity of anatase TiO_2 by coexposed {001} and {101} facets[J]. J. Am. Chem. Soc., 2014, 136:8839-8842.

[24] YANG H G, SUN C H, QIAO S Z, et al. Anatase TiO_2 single crystals with a large percentage of reactive facets[J]. Nature, 2008, 453:638-641.

[25] LIU N, LI K, LI X, et al. Crystallographic facet-induced toxicological responses by faceted titanium dioxide nanocrystals[J]. ACS Nano, 2016, 10:6062-6073.

[26] LU Y, ZHANG H, WANG H, et al. Humic acid mediated toxicity of faceted TiO_2 nanocrystals to daphnia magna [J]. J. Hazard. Mater., 2021, 416: 126112.

[27] FOO C, LI Y, LEBEDEV K, et al. Characterisation of oxygen defects and nitrogen impurities in TiO_2 photocatalysts using variable-temperature X-ray powder diffraction[J]. Nat. Commun., 2021, 12: 661.

[28] PAN J, LIU G, LU G Q, et al. On the true photoreactivity order of {001}, {010}, and {101} facets of anatase TiO_2 crystals[J]. Angew. Chem. Int.

Ed., 2011, 50: 2133-2137.

[29] WANG W K, CHEN J J, LI W W, et al. Synthesis of Pt-loaded self-interspersed anatase TiO$_2$ with a large fraction of {001} facets for efficient photocatalytic nitrobenzene degradation[J]. ACS Appl. Mater. Interfaces, 2015, 7: 20349-20359.

[30] YANG H G, LIU G, QIAO S Z, et al. Solvothermal synthesis and photoreactivity of anatase TiO$_2$ nanosheets with dominant {001} facets[J]. J. Am. Chem. Soc., 2009, 131: 4078-4083.

[31] LI A, WANG Z, YIN H, et al. Understanding the anatase-rutile phase junction in charge separation and transfer in a TiO$_2$ electrode for photoelectrochemical water splitting[J]. Chem. Sci., 2016, 7: 6076-6082.